Jürgen Schrempp

*and the Making of an
Auto Dynasty*

Jürgen Schrempp

and the Making of an Auto Dynasty

JÜRGEN GRÄSSLIN

MCGRAW-HILL

New York San Francisco Washington, D.C. Auckland Bogotá
Caracas Lisbon London Madrid Mexico City Milan
Montreal New Delhi San Juan Singapore
Sydney Tokyo Toronto

Library of Congress Cataloging-in-Publication Data

Grässlin, Jürgen
  [Jürgen E. Schrempp. English]
  Jürgen Schrempp and the making of an auto dynasty / Jürgen Grässlin.
    p. cm.
  Includes index.
  ISBN 0-07-135132-9 (cloth)
  1. Schrempp, Jürgen E. 2. Chief executive officers—Germany—Biography
  3. Daimler-Benz Aktiengesellschaft. 4. Chrysler Corporation. I. Title.

HD9710.G44 .D34313 2000
338.7'6292'092—dc21
[B]

                                          99-054060

McGraw-Hill

A Division of The McGraw·Hill Companies

Original German language edition: Copyright © 1998 by Droemersche
Verlagsanstalt, Th. Knaur Nachf., München. All rights reserved.

1 2 3 4 5 6 7 8 9 0 DOC/DOC 9 0 9 8 7 6 5 4 3 2 1 0 9

ISBN 0-07-135132-9

Set in Minion by Tina Thompson.

Printed and bound by R. R. Donnelley & Sons Company.

Dedicated to my parents,
Irmgard and Heinz

CONTENTS

I would be the wrong man for the job
if I couldn't accept criticism.
Jürgen E. Schrempp

Well practiced in the art of handling criticism, Jürgen E. Schrempp has refused to comment on this biography. The following dialogue, which took place in the author's presence at a September 18, 1998, press conference on the DaimlerChrysler merger at the shareholders' general meeting), is typical of his stance.

Sigrid Faltin, TV reporter: "What is your opinion of the profile Jürgen Grässlin has written about you?"

Schrempp: "Ask him yourself. I have nothing to say on the matter."

Faltin: "Why not?"

Schrempp (to the other journalists gathered in the press center): "Are there any other questions?"

Christoph Walther, manager of communications for DaimlerChrysler: "Okay, no more questions. We'll meet again soon."

The press conference ended there.

Freiburg, October 1999
Jürgen Grässlin

ACKNOWLEDGMENTS

First, my most heartfelt gratitude goes to my beloved wife, Eva, who supported this book project from the first line to the last and accompanied me to South Africa. Yet again our children, Sandra and Philipp, showed their full understanding about their father's spending far more time on the move, interviewing, and in the office than attending to their needs.

I would particularly like to thank Jürgen E. Schrempp, who through his openness, his readiness to talk, and his unflagging joy of discussion made this work possible—and many would surely be surprised by this personal interaction between such a man and this critical, pacifist, environmentally oriented author. His cooperative attitude in turn led to a number of interviews that would otherwise have been unlikely to take place.

Daimler's press spokesman, Christoph Walther, proved to be a highly enthusiastic discussion partner. Not only did I have the most meetings with him, but also the longest.

In South Africa, Christoph Köpke turned out to be an exceptionally fair individual whose persona is exemplary of the new, democratic, and open South Africa. I am grateful to MBSA secretary Wendy Hoffman for giving me access to a great amount of factual information and contacts, made through MBSA, which would hardly have been possible through my own research.

I would like to thank my friends for checking portions of the text and for their advice and encouragement. And I thank all those others—who could not be named—for providing information and material.

Finally, I would again like to give my special thanks to all those who provided information and answers in interviews. Without them this book could never have attained such quality and informative content.

PREFACE

This book was based on about one hundred meetings and interviews as well as the examination of a number of confidential documents obtained from Daimler-Benz itself. This meant that I could largely do without secondary sources and could instead let those actually involved have their say.

Jürgen E. Schrempp himself was interviewed primarily in sessions usually lasting about two hours—he had agreed to five such meetings in Stuttgart and Munich. I have also included statements from three previous internal meetings with the chairman of the board as well as various public appearances of his. The quotations published in this book, taken from conversations between the author and Jürgen E. Schrempp, have been authorized by Mr. Schrempp himself. In the case of all the other individuals I interviewed, authorization took place insofar that they themselves expressed a wish to be interviewed.

Particular attention was paid to interviewing Schrempp's former and present close friends and associates—including his personal consultants, Hartmut Schick and Detmar Grosse-Leege; the Group's spokesmen, Christoph Walther and Matthias Kleinert; and friends such as World Bank president James Wolfensohn.

I carried out interviews with Schrempp's closest confidants in four South African cities and the production works in East London; these form the basis of my account of his rise from a lowly member of the sales staff to chairman of MBSA. My interviews involved meetings with, among others, his friends, drivers, and secretaries; representatives of the automobile and supply industries; his farm administrator; the Dasa head there; and the current MBSA chairman, Christoph Köpke.

Interviews with the leader of negotiations for the employers' associations and the trade union NUMSA and its forerunner organizations, including South Africa's current deputy minister of labor, Les Kettledas, and German antiapartheid activists allowed me to analyze the business policies of the Mercedes board and its chairman during the time of racial segregation.

Friends of Schrempp, well-known business journalists, and qualified critics of Daimler's chairman, in particular trade unionists and the lawyer Martine Dornier-Tiefenthaler, were all given their say in this profile. Ecological considerations were discussed with environmental expert Werner Pollmann, with the head of the Wuppertal Institute, Professor Ernst Ulrich von Weizsäcker, and with Alexander Dauensteiner, spokesman for the umbrella organization of critical Daimler shareholders.

In questions concerning the man, Jürgen E. Schrempp, and his path through life, meetings with his brothers, Günter and Wolfgang, were helpful in allowing me to include biographical information.

Interviews with several management and supervisory board members took place to allow presentation of the business policies at the management level. Former chairman of the board Werner Breitschwerdt shed light on the history and certain current aspects of the company.

Edzard Reuter and Helmut Werner understandably declined to grant interviews. Wolfgang Piller and Werner Heinzmann, of the Dasa board of management, canceled previously scheduled meetings at short notice or refused to attend them on formal grounds.

A not inconsiderable number of people I interviewed feared the negative consequences arising from even relatively harmless opinions and comments being quoted in this book. But because I felt that their statements and content were important, I have in various cases included the quotations without attributing them by name. In particularly explosive cases I have also suppressed the gender of the speaker to be absolutely sure that there could be no possible reprisals. As the author of this work, I vouch for the accurate reproduction of such quotations.

This manager profile was published in the German-speaking world at the beginning of August 1998. Only a few weeks later it had climbed onto the nonfiction best-seller lists. It was in the top ranks of the best-seller list of the famous Hamburg magazine *Der Spiegel* for more than nine months. The weekly magazines *Focus* and *Wirtschaftswoche* gave it an excellent second place on their lists of best-selling business books.

This portrait of Schrempp appeared in the People's Republic of China in translation in spring 1999.

This edition for the English-speaking world has been reworked by the author; current developments up to the shareholders' general meeting on May 18, 1999, have been included.

Freiburg, October 1999
Jürgen Grässlin

INTRODUCTION

An Unusual Type: Ability, Competence, and Ambition

We saw his qualities, we saw his strengths, we saw his visions.
Theo Swart, Managing Director, McCarthy Group Ltd.

Conventional business biographies tend to share a particular tone—benevolent, respectful, and occasionally glorifying. What they do universally is celebrate the subject's success in a way that is reminiscent of elite portrait painting in the Middle Ages, when the beauty or grace of a noble person was exaggerated as an act of homage.

Consider how such a portrait of DaimlerChrysler's cochairman might begin:

———

Jürgen E. Schrempp's rise from a lowly car mechanic at a branch of Mercedes in southern Germany to one of the most powerful men in the world is the result not of coincidence or luck, but of skill and competence. The route that took Schrempp from a simple background through South Africa, the United States, and Munich to the eleventh floor of Daimler-Benz's headquarters in Möhringen, a suburb of Stuttgart, was long and hard, but the charismatic Schrempp's character traits, abilities, and skills allowed him to make his way to the top of Europe's largest company.

Schrempp made no secret of the great ambitions he harbored. "I believe that I will reach a position of leadership," he told Leo Borman, one of the leading figures in South Africa's automobile industry. "I, Jürgen Schrempp, believe in Jürgen Schrempp."

Gerhard Liener, who as director of mergers and acquisitions at the Daimler-Benz Group's headquarters in Germany was responsible for overseeing the South African subsidiary in which Schrempp rose from technician to top manager, recognized Schrempp's talent and declared that he would become chairman of Mercedes-Benz South Africa. During the 1970s and 1980s Theo Swart, chairman of the influential McCarthy Group, the leading dealer of vehicles in South Africa, and one of the Cape's leading industrialists at that time, dealt frequently with him and recognized even then that Jürgen Schrempp would scale the heights; he says he had no doubts that Schrempp would someday be a major figure at Daimler-Benz headquarters. In this he was in full agreement with Morris Shenker, Schrempp's predecessor as chairman of Mercedes-Benz of South Africa (MBSA). Schrempp, too, concurred: "I will only return as chairman [of MBSA]," he confided to his private secretary, Waltraut Lenhard, brimming with confidence in himself and his abilities. Such statements are born of a desire for power.

Schrempp has a high intelligence quotient, says Leo Borman. This was one of the factors that paved the way for his climb to the top. Furthermore, Borman believes that Schrempp's particular strength is that he doesn't allow himself to be "influenced by EQ, the emotional quotient." This is "a strong trait" that Borman, the former head of one of South Africa's largest automobile companies, recognized in Schrempp early on, telling this rising star, "One day you will be chairman of Daimler-Benz, not just Mercedes." So Borman was not surprised to receive a phone call from Schrempp just after he had been elevated to head of Daimler-Benz. Looking back, the seventy-six-year-old Borman cites one other factor that he believes may have played a role in Schrempp's rise: "Schrempp is a Virgo, and that signifies strength."

Some have called Schrempp's strength ruthlessness, citing the massive layoffs that occurred under his leadership in the 1990s. But how does one judge a business executive's true character?

Perhaps just by looking at the figures for sales and operating profits?

Exactly, some would say. Others would counter that it is necessary to look more closely at the man's human characteristics. Certainly the massive lay-offs under his stewardship brought Schrempp neither pleasure nor satisfaction. And anyone who fails to accept that today's economic necessities cannot be compared with those of ten years ago is still living in the age before globalization.

Richard Wentzel, a happily married father of three who is one of the best-known professional drivers in Cape Town and who has driven everyone from ambassadors to soccer stars, has seen how Schrempp handles the little things. In 1996 Wentzel became Mercedes-Benz South Africa's sole company driver; now he takes company guests to their destinations, and at least once or twice a year chauffeurs the boss of bosses himself.

Whenever Schrempp comes to Cape Town it is Wentzel who anticipates his every wish. Sometimes an excursion in the Mercedes-Benz 600 limousine takes them to one of the numerous wineries or to the waterfront. Another time they use the Mercedes-Benz 420C to go to Cape Point. "Something different every day" is Schrempp's motto—he is not the type to frequent a favorite restaurant, for example. Repetition would be monotonous.

Wentzel recalls that one day he and Schrempp were headed east for a Pavarotti concert at the sports stadium in Stellenbosch, South Africa's second oldest city, which has made a name for itself as much for its wine as for its student population. After collecting Schrempp's business friend Anton Rupert, one of South Africa's leading industrialists, Wentzel noticed that they were running late, and asked Schrempp, "Shall I drive over the speed limit?" Schrempp nodded, and the three of them reached Stellenbosch in record time. Once there, Wentzel assumed he would be in for a long evening of waiting—which was no reason to complain, from his point of view, as endless waiting is part of the job. Schrempp said good-bye and strutted off with Rupert, only to return a few minutes later with an excellent supper for Wentzel. "My driver must eat," he insisted to the thoroughly nonplussed South African.

Wentzel can still remember well their very first meeting, in December 1995. Wentzel collected the world-famous company head from the airport and took him to his destination—where he was immediately invited in by Schrempp and his wife for a visit. The driver was impressed by that, but then Schrempp added, "Now I'll listen to you," and sat back, allowing

Wentzel to both entertain and inform him. That is yet another aspect of Schrempp—he is a man who knows how to obtain information, whether by phone from the president of the World Bank or over a drink with a friendly driver. And it reveals a side of Schrempp antithetical to his reputation for being hard and uncompromising.

The night of the Pavarotti concert, Schrempp did not return to Cape Town again until very late, but Wentzel was told to be ready to drive at six-thirty the following morning. Also typical for Schrempp was that the next day he did not let himself be waited on, opening the door himself and putting his own luggage into the trunk. Schrempp lit his first cigarette during the drive to the early-morning meeting and started his working day in the backseat of the company car, making phone call after phone call, even at that early hour. Wentzel does not understand German, but he was able to pick up the meaning from Schrempp's tone of voice—and it was clear that some of the people Schrempp spoke to were getting a good dressing-down. The manager in him had reclaimed Jürgen E. Schrempp.

———————

The present chairman of MBSA is without doubt one of Schrempp's closest friends. Schrempp himself promoted Christoph Köpke to the top of the South African company as a gesture of friendly appreciation. The relationship between the two is one of complete frankness—Köpke is one of the few people who tells Schrempp exactly what he thinks, both the negative and the positive. This is true not only of their frequent business meetings but also particularly of their shared farm in Eastern Transvaal.

Köpke, who moved from Germany to Cape Town with his parents when he was just four, has known Schrempp ever since Schrempp began working for MBSA. Köpke's honesty and his contentment with what he has achieved guarantee that his praise of, and respect for, Schrempp have nothing whatsoever to do with promoting his own career or business prospects. Looking back, Köpke explains Schrempp's success on the basis of one unique ability: He was able to encourage his fellow workers in the sales departments of United Car and Diesel Distribution Pty. (UCDD) and then MBSA through the difficult times as almost no one else could. "Jürgen Schrempp has mastered the art of motivation," Köpke says of his predecessor, though Köpke is referring to situations and times when things were, on the whole, going relatively well. This talent earned Schrempp respect and support from many quarters; his employees were all prepared to walk through fire for their boss.

As someone who knows Schrempp's family relationships well, Köpke also highlights the immense influence of Schrempp's wife, Renate, who he says always managed to give her husband a good scolding at exactly the right time. "He has enormous respect for Renate," Köpke points out, and adds that it is largely thanks to her positive influence that this powerful manager has not succumbed to the arrogance of power but has hung on to his humanity. On the other hand, says his friend, what doesn't suit Schrempp so well is that "he doesn't have an awful lot of say at home." The notion that there is a strong woman behind every successful man may be a cliché, but it seems to be true for Schrempp.

Detmar Grosse-Leege knows Jürgen Schrempp like almost no other. For many years Grosse-Leege, a media professional who was once elected PR manager of the year, has served as Schrempp's press officer and personal advisor. This is what he says when he is asked to sum up his boss in a few words: "His fairness, spontaneity, openness, decisiveness, and strategic far-sightedness are what impress me." Schrempp "expects a lot from his employees; he demands precision, tolerates contradiction, and loves constructive criticism."

When Werner Breitschwerdt, a former chairman of the board, is asked whether Schrempp is a superman, he answers, "He wouldn't be where he is now if he weren't."

But it is also the case that without luck and the support of influential people, Schrempp would probably have never—or at least not at such a relatively young age—reached the top of the Daimler-Benz Group. Among his few (and therefore conspicuous) weaknesses is an informality unusual for a Daimler chairman, and an unconventional streak that he used to find difficult to restrain. For what is normally considered to be a human strength quickly becomes a career hindrance difficult to compensate for in the highly selective business world.

And while his informality is still a distinguishing feature, his uncon-ventionality has long been brought under control. Many years have passed since Schrempp, in an apparently arrogant moment, spoke of how Daimler-Benz needed him but he did not need Daimler-Benz. And he has outgrown escapades such as that on the Spanish Steps in Rome, where he was stopped by police in the middle of the night while out with two close associates; the story was distorted and blown up out of all proportion in media reports.

Without doubt, Schrempp will have to continue to give his full atten-tion to radically restructuring the Daimler-Benz Group, maintaining a for-ward-looking portfolio and creating a socially acceptable profit orientation. He will have to be careful not to walk into any of the traps set by his ene-mies, who think of him as a sort of Rambo and who cannot wait to see him make a mistake. He will certainly have to take care if he wants to continue to master his weaknesses. And even today much of the public continues to perceive him as not just one of the most responsible and assertive managers in the world but also as one of the most ruthless and power-hungry. He will have to keep working on this aspect.

When Jürgen Schrempp took over the leadership of Europe's largest indus-trial enterprise in May 1995 his predecessors had left him a company in which there was a lot of diversification and very little integration. The dis-astrous balance sheets inherited from Edzard Reuter's spell at the top were the worst that a European corporation ever had to contend with.

Yet Schrempp needed only three years to sell off the loss makers, to knock into shape the twenty-three divisions that remained after the portfo-lio had been consolidated, and to achieve profits the like of which had never been seen before. In typically understated fashion, Schrempp said that Daimler had concluded 1997 "with great success" and that all activities had become "more successful and more profitable" thanks to "our concept for increasing the value of the company." The figures for summer 1998 speak for themselves: sales up by 19 percent to $66 billion and an operating profit up by 79 percent to $2.3 billion. No German manager before him had ever been in a position to make such an announcement.

Gone were the days when the nightmare of a ruthlessly high shareholder value would make union leaders go to the barricades and bring the work-force out onto the streets. The radical restructurer Schrempp restored social peace by creating twelve thousand new jobs in one year alone. He has brought a breath of fresh air to the once-stodgy product line: 80 percent of all products are less than five years old. The more than 715,000 Mercedes cars that were sold in 1997, 11 percent up on the year before, are witness to an unprecedented boom.

"We have unleashed a new entrepreneurial spirit throughout the whole group," announced Schrempp contentedly at the beginning of 1998. At Daimler-Benz work has become, in Schrempp's words, "a lot more fun" for everyone—including the board of directors.

A highly motivated team, controlled by one man with a certain amount of luck and a lot of skill, is responsible for the progress that not long ago would hardly have been thought possible. But Schrempp, whose nickname, "Lord of the Stars," goes along with the chairmanship of Mercedes-Benz, has not yet achieved his objective. He wants to take the Daimler-Benz Group to the very top: to make it the world's most modern and lucrative business, number one among all mobility and service groups. It is not long ago that Schrempp's sales target for 2008 was the apparently fabulous figure of $133 billion. But what seemed so impossible only a short time ago has happened sooner than even he predicted, as that target has already been exceeded thanks to the 1998 merger with Chrysler. And regardless of where Schrempp's journey will take him next, Daimler-Benz's star is shining brighter than ever with him in the vanguard.

Again and again Schrempp finds himself faced with a problem he cannot immediately solve. But he prepares himself perfectly for the task ahead, finds a way—perhaps one not even considered by others—and finally achieves the impossible.

When he moved to South Africa he found a somewhat reserved envi-ronment in which he succeeded in climbing from an insignificant technical services manager to become chairman of MBSA and giving new sheen to South Africa's Mercedes star. With tact and sensitivity he even mastered the tricky political situation of producing luxury limousines for the white rul-

ing classes during the apartheid era. In the 1980s he shifted his attention to Euclid Inc. in Cleveland, where he was supposed to restructure an ailing subsidiary—and instead sold it off profitably.

His biggest challenge, however, was when he had to form Deutsche Aerospace (Dasa) from a conglomeration of oversized and splintered German aviation and aerospace industries dominated by internal interests. Today's Daimler-Benz Aerospace has Jürgen Schrempp to thank for its leading position among the world's largest aerospace companies. This achievement is all the more impressive given the disastrous background (the collapse of the Communist bloc and an extremely weak U.S. dollar) against which Schrempp had to act. Today, Daimler-Benz Aerospace can boast of a glut of orders and a positive balance sheet, even if this did take some time to achieve. It was Schrempp who, during his six-year spell as head of Dasa, laid the foundations that now allow the company to be in the black and look confidently to the future.

In the summer of 1995, following the failure of the visionary Edzard Reuter, Schrempp became chairman of Daimler-Benz's board of directors. What he has since achieved speaks for itself. Under his leadership, Germany's showpiece company has developed into one of the world's most innovative and profitable businesses. No challenge is too great for Jürgen Schrempp, it appears. And since the fall of 1998 he has led, together with Robert J. Eaton, the DaimlerChrysler Group. Those who have reached the top know how easy it is to fall from grace, but in Jürgen Schrempp's case, his competence, ability, and great ambition will protect the Lord of the Stars from this danger. . . .

The pattern of a typical corporate biography is this: describe and admire, analyze and worship, and, where applicable, criticize for the sake of form. That will not be the case here. These introductory passages do indeed reflect one side of the reality of Jürgen Schrempp as it is perceived by many. A large majority of those I interviewed were generally positive about Schrempp. Why this should be so will be looked at in more detail later.

But that is just one side of the coin. The other is that I got the impression, right from the start of my work on this book, that a not inconsiderable number of people wanted to provide me with a certain type of information, best suited for a standard business profile. But from the very beginning of this project, questions and contradictions cropped up. The more information I collected about Schrempp, from a surprising range of sources, the more I was able to build a solid, critical picture of the man and the manager.

That this venture succeeded at all is due to the people I met, both within and outside the company, who shared my interest in presenting as objective a view as possible of the individual at the company's head.

Of course Daimler-Benz is looking better now than at any time in its history, and of course Jürgen E. Schrempp is to a large extent responsible for this progress. His image among the German public—sometimes significantly less than positive—has noticeably improved in tandem with the company's balance sheet, and internationally he has long been ranked at the top. Of course Schrempp, the manager, displays numerous outstanding abilities and traits (which was why I chose to examine the man at the top of the Daimler-Benz Group so intensively), and of course in many ways he stands out favorably when compared with the vast majority of his colleagues on the boards of large German businesses—not least for his sense of humor, which explains why some passages of this book are formulated so ironically. Schrempp's less well known stumbling blocks, without which no portrait would be complete, will also be presented later.

The basic problem for biographers is that they are generally dependent on the object of their desire, so to speak. Anyone who wanted to write about Edzard Reuter, for example, had to tread a conformist line to obtain background information. Rejection or collaboration, refusal or cooperation—these turn a work such as this into either an analysis from the outside or a glance behind the scenes. This book does not claim to be a comprehensive biography; that could be written only ten or twenty years from now. Jürgen Schrempp is still thirsty for action and is good for another book's worth of surprises. This book is simply intended to present his economic, social, and personal views as well as map out his career and attempt a critical evaluation.

Many doors were opened to me because of the commendable readiness on the part of Schrempp, and at least some of his friends and colleagues, to take part in conversations and dialogues, and it really was possible to gain a glimpse behind those thriving scenes of sales success, share value, and profit. The Lord of the Stars proved surprisingly willing to talk to me, an author who is known to be among the company's critics. No other chairman of the board has ever dared to speak this freely during the hundred-year history of Germany's showpiece company. For this, Schrempp has earned my thanks.

However, it must be said that his willingness to cooperate was not without limits. Even if he was prepared to talk, he categorically refused to allow access to internal company documents—even when some of these had already been made public by the media. I was therefore forced to obtain this information by other means.

When I first began carrying out interviews and background conversations, I did not have any clear idea of what the result of more than one and a half years of research would be. Almost every new interview, telephone call, letter, and fax contributed another piece of the puzzle. The result is a book presenting Jürgen E. Schrempp as I experienced him in face-to-face conversation and during his public appearances. The many interviews with him, and the numerous other interviews I conducted with all sorts of people, allowed me to look deeply into the inner life of the company in general and Jürgen Schrempp in particular. This has resulted in a far more complex and differentiated picture of Schrempp, the man and manager, than has been suggested up to now.

That I was never interested in presenting a biography of a hero may not always suit the man portrayed here. My aim, however, has the advantage of at least coming close to offering an objective picture. And if the subject described in this book feels immeasurably misunderstood and misinterpreted, then he of course is free to look for a more respectful biographer—one who would straighten everything out.

PART 1

The Careerist

CHAPTER 1

The Hyena

The softest thing about him are his teeth.
Hugh Murray, Publisher

"One cannot simply stay at home in Germany if one wants to have a successful career in an international group"—that is an article of faith for the dynamic Schrempp.

But the idea that he would one day have such a career probably never crossed the mind of anyone who knew young Jürgen during his school years.

"I got stuck in my third year," Schrempp admits now, grinning, for he has long since gotten over his embarrassment at having to repeat a year at Freiburg's Rotteck Gymnasium (the close equivalent to the American high school), where his attitude could only rarely be brought into line with the expectations of the majority of the teachers there. The natural sciences were the only subjects for which he displayed any talent, and it is no coincidence that the one teacher whose name he can recall was his mathematics/physics teacher; the others he describes as "meaningless" and claims they were "unable to motivate me." The young Jürgen was simply not prepared to learn what he saw as irrelevant facts, and he was particularly weak in languages, categorically refusing to learn even the basics of vocabulary and grammar.

Schrempp still seems to have an aversion to paper, except for the small notepad he always has on him. Otherwise almost everything in the life of Jürgen E. Schrempp has changed—even his attitude toward foreign languages, including English. Today Schrempp can prepare his own drafts, give off-the-cuff speeches in fluent English, and smoothly switch from one language to the other in order to choose exactly the right word.

But what Schrempp's school career shows is that even at a young age he could not be forced to fit into any particular system against his will. At the time, no one likely imagined that this extremely intelligent odd-ball, who so stubbornly resisted learning, would become the most powerful business force in Germany. Indeed, Schrempp left school early; his brother Günter, two years younger than Jürgen, recalls Jürgen confidently coming out with the statement "My plans do not involve university entrance exams." And so Jürgen Schrempp began his apprenticeship at a local branch of Mercedes-Benz.

Schrempp liked nothing better than fiddling around with vehicles, and even in his early years he gained a reputation as a skilled specialist in trucks. "That was the hardest job," says Günter Schrempp with respect, particularly as Jürgen has not lost any of his expertise and can still repair vehicles with the best of them.

As a youth, Schrempp could not know that he would spend the next few decades working for Daimler-Benz. But one critical relationship early in his career set him on the path that would eventually take him to the top of the very same company at which he started out as a mechanic. Karlfried Nordmann, manager of the Freiburg branch of Mercedes-Benz, was a former military man who quickly became aware of the young apprentice—though less for his considerable skills as a mechanic than for his musical talents. Fighter pilots from Nordmann's old squadron used to meet regularly for reunions at this Mercedes branch, and Schrempp would play an emotive rendition of taps on his trumpet for the fliers. Eventually Schrempp and Nordmann struck up a friendship, and it would be Nordmann who planted in Schrempp the notion that he did not need to spend his rest of his life moldering away in this southern German province.

Even though he had left school early, Schrempp had had no trouble passing his secondary-school exams, and on the basis of this he was admitted to the school of engineering in Offenburg, an industrial city in central Baden.

Schrempp enjoys looking back to his "lovely time" there, which ended with his receiving a certificate in engineering.

At that time Schrempp was low on funds: "I was not wealthy and had to earn my money playing dance music," the millionaire of today says with pride. He formed a band with some friends and played at weddings, birthday parties, and festivals in his hometown. Schrempp had always found the trumpet—the classic lead instrument of bands in those days—particularly well suited to him, and though a fair number of things went wrong early on in his musical career, he was not the type to give up if something did not go well; that sort of experience would later prove useful for him in business.

After he finished his studies, Schrempp returned to Mercedes-Benz, where he was promoted without his having to do much. Part of this was the result of a technicality: When the school of engineering he attended in Offenburg was elevated to the rank of technical college, his engineering certificate was automatically upgraded to a regular degree in engineering.

A year and a half later, Nordmann became interim manager of the Berlin Mercedes branch, and shortly afterward he was given the position of manager at the Mercedes branch in Hamburg. Schrempp was captivated by the older man's cosmopolitanism, and began to develop an interest in getting a job outside the country.

Here Nordmann was able to open the door for Schrempp.

The two of them transferred to Daimler's headquarters, then located in Untertürkheim, at almost the same time, and Schrempp was put in charge of engineering for trucks and buses at what is now the central customer service department. This marks the real beginning of his career.

Schrempp understands that he must let his body recuperate from the long periods of stress he endures in his job. His annual summer mountain climbing expeditions are simply a form of "preventive medicine," according to Reinhold Messner, considered one of Europe's most experienced mountaineers. Every year Schrempp is on the team when a small group of managers attacks the Alps in summer. During these expeditions, which last several days, Schrempp "is a good comrade," reports Herbert Henzler, a friend from Freiburg who is with the German branch of the American consulting company McKinsey. "There is no trace of Mr. Mercedes-Benz while climbing or when we make camp."

Messner says that Schrempp simply "enjoys going up cliffs," but the achievement and the risk-taking are equally characteristic of Schrempp. Schrempp makes intensive preparations for the trip and is, according to Messner, more powerful than the rest of the team. "He is crafty enough to start training in plenty of time and make the necessary lifestyle adjustments before the tour," including cutting back on his tobacco and alcohol consumption, which can affect one's fitness. Or at least that's what he does now. One year Schrempp and Messner, whose birthdays are two days apart, decided to celebrate by climbing twelve-thousand-foot Mt. Ortler. Henzler says that Schrempp "had a moment of truth on Ortler."

"Ortler is a relatively difficult mountain, not without danger, with some steep sections," says Messner, and he notes that climbers have to be careful not to overestimate their ability. But this time Schrempp evidently thought he was more fit than he really was, and he wound up dangling precariously over a 150-foot drop as he waited for their guide to rescue him.

"The situation was life-threatening," remembers Henzler.

Afterward Schrempp swore that he would never smoke again—though he did not manage to keep this pledge for very long.

That episode was apparently something of an aberration. When ascending a rocky cliff, Schrempp shows absolutely no uncertainty or even nervousness. His rapid ascents and descents and the naturalness with which he climbs have earned him the nickname "Cliff-face Schrempp." "He doesn't cut a bad figure," says Messner. "He radiates strength and success, not worry."

Karlfried Nordmann became a divisional manager and head of the domestic and foreign central customer service department at the beginning of September 1968. Schrempp took on the responsibility for looking into export problems involving trucks and buses; he was anxious to move on, but he stayed for the time being for the sake of his boss. When Nordmann, who has since become a director, was promoted to president of Mercedes-Benz USA in January 1971, it also became clear to Schrempp that he wanted to go abroad.

Schrempp's brother remembers the situation well: "Jürgen's dream was to go to the States." But, Günter recalls, everything turned out differently, because "then came this offer from South Africa."

The situation Schrempp found in South Africa was a complicated one. In 1958 the South African Car Distributors Assembly (CDA) had signed a contract to assemble Mercedes products. Four years later the South African Auto Union acquired the franchise for Mercedes-Benz and the Pretoria-based United Car and Diesel Distribution Ltd. (UCDD). After another four years UCDD took over CDA as a subsidiary.

Not too long after this, the South African government introduced a local-content law requiring that 66 percent of a car's weight be the result of local production. This essentially forced foreign companies to manufacture most of their product in South Africa. Mercedes consequently increased its investment in the area, acquiring a considerable number of UCDD shares, and began producing S-class cars, meant for sale in South Africa, at the former CDA plant in East London, where trucks, vans, buses, and tractors were also made.

The company sailed close to the limits of South African legality as it tried to find ways of dealing with the government's rigid regulations. "Mercedes reacted quickly by producing the heaviest vehicles in the world," explains Ekkehart Friederichs, the present public relations head at the East London factory. Thicker steel was used intentionally, to increase the weight. This resulted in a South African Mercedes that on average weighed 330 pounds more than its German counterpart, and permitted it to meet the local-content requirement by using more South African steel. "This trick," says Friederichs, "allowed Mercedes to make local content as cheap as possible." (Since the end of the 1980s, however, calculations have been carried out according to value, not weight. As a result, says the PR man with a smile, "Mercedes no longer produces the heaviest automobiles but the most expensive.")

At this time there was a permanent state of conflict between Morris Shenker of UCDD and Leo Borman of CDA. The greatest similarity between the two executives was in their firm rejection of union organizations. Borman was considered a "difficult man," neither able nor willing to understand the employees' position. Union leaders, for example, were not permitted to solicit new members in front of the factory gates; they had to stay across the street. But despite this commonality, the hostility between the two hard-liners was so obvious that even Shenker's curriculum vitae mentioned the friction between the two, referring vaguely to "personal clashes and other obstacles."

Shenker began to worry about a decline in the quality of vehicle production when the local-content rules were introduced, and he realized that the conflict between him and Borman was starting to get out of control. Over the succeeding years the adversaries made attempts to cooperate and to increase contact between UCDD and CDA, but there were continual flare-ups.

When Schrempp first arrived in South Africa, he evidently misjudged the situation, not taking the running battle between the two executives seriously enough. But that wasn't Schrempp's only mistake early on. The autocratic, conservatively run automobile company had great reservations about this relatively aggressive youth. "What does he know that we don't?" longtime Mercedes employees asked spitefully about the new arrival from distant Germany. Complicating matters was that the thirty-year-old Schrempp had been in South Africa for only seven months before he got into a physical confrontation with a colleague from the customer service department, Karl Siebenrock. Besides leaving him with a few bruises, the clash did nothing but strengthen his reputation as an outsider.

But after a difficult initial period, Schrempp settled in, almost intuitively learning "to hold back when necessary" and "to take the offensive when advantageous," according to Christoph Köpke. Fred Dill, the service manager, was Schrempp's first boss—and he stood in Schrempp's way.

"Schrempp shelved him," Johan Frederik van Olst, a secretary in MBSA's marketing planning department, notes dryly. Schrempp worked things so that he was in charge of the more profitable aspects of the customer service department, and Dill wound up with all the flops. The flamboyant Schrempp found it easy to nudge out his superior, particularly as Dill was seen as humorless and lacking any kind of managerial ability, and soon Schrempp was the manager of the customer service department; later he became the national service manager.

In customer service, managers have a wide range of duties, including training the team, providing service, organizing maintenance, and developing new guidelines. Schrempp had to visit more than 120 privately owned Mercedes branches scattered around the country, and he spent a considerable amount of time flying around a country that is five times as large as West Germany was before its reunification with the East.

Schrempp cannot have found his job particularly pleasant. If a Mercedes-Benz customer was dissatisfied, the dealer, or sometimes even Schrempp himself, would receive the comments, and ultimately angry customers' letters of complaint would land on Schrempp's desk. He had to force the technicians at Mercedes' predecessor, CDA, to adopt his company's quality standards, and it was also his job to negotiate with the dealers.

Another problem Schrempp encountered early on was that in the automobile division, customer service and parts managers are looked down on, treated as second-class citizens. But here Schrempp displayed one of the characteristics that would serve him well in the future. Instead of complaining about the situation, Schrempp turned it around, forming his staff into a team and suggesting to them that they were in fact the ones that were on top and that the others were second-class. This immediately improved the work environment and created a cadre of employees who were devoted to Schrempp.

What this episode also reveals, according to one observer with years of firsthand experience of Schrempp's patterns of behavior, is that you are either with Schrempp or against him. If you choose to support him, this source says, then "he will defend you beyond the bounds of reasonableness, beyond every normal expectation." But if you oppose him, "be careful; he will demolish you."

Hard on Himself and Others

I could have killed him.
Karola Block, Secretary to the Management Board

He was merciless.
Johan Frederik van Olst, Marketing Planning Department of MBSA

The highway through the Kalahari Desert leads to the Okavango Basin in the north of Botswana. After stopping for gas in Schilpad's Gate, near the South Africa–Botswana border, and for beer in Lobatse, a few miles inside Botswana, Jürgen Schrempp and Christoph Köpke were cheerful as they drove their off-road vehicles north in one of the world's most beautiful natural settings. The two friends had toured Botswana before, with their families, two years earlier, in 1979.

But this time catastrophe struck without warning. A moment's inattention, and Schrempp lost control of his vehicle, which left the road and crashed into a rock. The car, in which his wife, Renate, and son Marc were passengers, overturned, dragging the trailer with it.

Renate and Marc were not seriously injured, but Schrempp himself was bleeding from a deep arm wound.

Köpke crammed the Schrempps in his own car and sped to Lobatse, the nearest town with a hospital, only to find that the doctor on duty was drunk. Nevertheless, Schrempp's arm needed to be stitched up, and so he was placed on one of those old operating tables with cast iron feet and given thirty-six unsteady stitches. Despite his injury, Schrempp refused to stay in the hospital, though he had lost quite a lot of blood. Instead, he gave himself antibiotics and allowed himself much too short a period of rest in a hotel room before he decided to continue with the trip. Perhaps one of the things urging him to get back on the road was his anger at his own carelessness, which was gnawing at him. These things shouldn't be allowed to happen—at least not to him.

Schrempp climbed back into the driver's seat, but he and Köpke decided to change direction, aiming for Botswana's capital, Gaborone.

Schrempp was not well, however, no matter how much he insisted he was; even though the temperature outside was 104 degrees, he began shivering.

At night, after the others in the group put up their tents, Schrempp drank a quantity of brandy and pulled out the stitches from his arm himself—his upper right arm still bears the scars. Both families bore the brunt of his foul temper; the thoroughly negative attitude he had developed toward the journey spoiled it for the rest of them as well. The problem, however, was within himself: He felt like a cripple for the rest of the journey, incapable of even putting up a tent. This is typical of Schrempp, says Christoph Köpke years later: "He goes through with things despite there being overwhelming negative factors."

Jürgen Schrempp is a fighting type who would rather go to work sick than stay home. He has only been off work once because of sickness, according to his secretary, when he spent one day in the hospital and two more at home because of an operation on his tonsils; even then, he returned to work far too early, collapsed in the office, and had to be taken home. On that occasion, said his secretary, "he looked like death warmed over."

Schrempp is the sort of individual who is constantly testing his physical and mental limits—and he expects the same from his fellow workers. He is hard on himself and others.

Some coworkers adapt better to Schrempp's demands than others. Waltraut Lenhard, Schrempp's former secretary, herself enjoys a fast-paced atmosphere, and counts the years she worked for Schrempp among the best of her career with MBSA. "He needs everything yesterday," she says of her former boss. And she adds, "He's a man who needs problems—otherwise he'll create them himself." When she finally left, Schrempp presented her with a signed photograph thanking her, in his own ironic way, for the "many stormy hours" she had spent in his office.

Schrempp's next secretary, however, was not especially thrilled when she learned that Schrempp had asked that she be assigned to him. Karola Block had no interest in leaving her current boss, Erich Glanz, one of Schrempp's colleagues, and argued against the transfer.

Her first day in Schrempp's office left her with a strongly negative impression. Schrempp came in with a briefcase that, Block remembers, "was as large as a child-sized coffin," and he helped himself to her cigarettes. Later she went, under protest, to buy him his own pack of cigarettes, but he never paid her back for them. At the end of that day, she recalls, she thought the man was "very strange."

Ultimately, however, Block, who was previously responsible for truck acquisitions in South Africa, would spend more than two and a half years with him. During this period she would often have to stand up resolutely for herself—otherwise she would have been hopelessly lost.

Schrempp is still driven by an incessant inner turbulence, whether with colleagues or friends in a restaurant, walking in a beautiful Black Forest valley, or on the eleventh floor at the Daimler-Benz headquarters. He is incapable of sitting still in a chair, as if he were being tormented by ants; he does not walk normally but struts through the countryside, or lights up one Marlboro after another (he keeps a second pack ready just in case). He switches the cigarette nervously from hand to hand. Meanwhile, the chain smoker discusses problems and profits, jobs and share values, as if his life depended on them.

This nervousness is nothing new. Waltraut Lenhard can still remember how "the whole table would shake" and how she had to keep going out to buy cigarettes for her boss. Lenhard's successor, Karola Block, also recognizes these symptoms; in her experience, he "cannot even sit still while he is dictating." But Block developed a strategy for dealing with it: "Give him a rocking chair."

When Block was working for Schrempp, his motto, "Go, go, go," plagued her "from morning till evening." "The stress really started piling up" with this new boss. "He thinks that there is nothing that cannot be done," maintains this dynamic woman. He brought out "the best" in her, and simultaneously "the worst."

Karola Block describes her ambivalence and awkwardness in the face of Schrempp's self-confidence. One day, Block recalls, he wanted to fly to Germany—on that same day, and it had to be with Lufthansa. Block found herself in a quandary, as there were no first-class seats left. She tried to explain the situation to the Mercedes chairman, but he put her down brusquely: "Girl," Schrempp said threateningly, "just let me know if you cannot do your job." He was saying, in effect, "Look for a quieter job if you cannot handle this one."

She has never forgotten this threat. That day she slunk out of the office "like a dog with its tail between its legs" and thought, "This is never going to happen to me again—even if he wants a pencil made of gold." And as it turned out, she was able to get him what he wanted. She made contact with the director of Lufthansa "by a roundabout route" and worked on him until she got what she needed. Schrempp was content, and Karola Block had learned her lesson: "The important thing is to get it done," and not "how it is done, what is done, or who I have to shoot to get it done."

High-energy Schrempp is blessed with the constitution of a horse. During the South African years he got an average of only three to four hours' sleep a night, and that had to suffice to recharge his batteries for his mercilessly full calendar. He would schedule meeting after meeting with only a short rest on board a plane. South African friends such as Hugh Murray, himself an excellent sportsman, consider the German "crazy" for the way he continually pushes his limits.

Even when Murray, now publisher of *Leadership* magazine, thinks back to the days when he would regularly go parachuting, he admits, "I was not in his league." Schrempp was physically fitter. The two of them would hit the bottle, sometimes all night long, during the flights Schrempp shared with his business journalist friend. Schrempp would keep "all the first-class flight attendants busy," and still be in good enough shape early the next morning to give a brilliant speech before a distinguished audience.

In South Africa Schrempp earned a reputation as a flamboyant, alcohol-proof speaker with staying power and a talent for rhetoric. These traits might seem to be worlds apart, but both are characteristic of Schrempp.

Gerd Andreas, a manager in Cape Town, still remembers a conference in Frankfurt for the automobile dealers. Schrempp partied with the dealers till four o'clock in the morning, "smoking and drinking like a lunatic," and was still able to give his usual high-quality performance at the lectern after breakfast. Drinking doesn't affect him, according to Waltraut Lenhard: He "was always at his desk afterward, and never had a hangover."

Typically, Schrempp, an early riser, would whirl into his office in Pretoria at seven in the morning and be the last one out of the deserted headquarters at about nine in the evening. "He was the first to arrive and the last to leave," confirms Karola Block, who had to adapt to her boss' habits. Instead of starting at half past eight, she would have to be ready for action three-quarters of an hour earlier. Once, when Block arrived at 8:15 A.M. after a dentist appointment, Schrempp gruffly inquired why she was late. And another time, when she had to leave at five o'clock to look after her children, he complained about the damage to the company caused by her early departure, and about the work that still had to be done. It was not unusual for Block to stay two hours longer in the evening after all the other secretaries had gone home.

There is more to this than just a workaholic and his insatiable appetite for work. When normal working hours are over he tours the offices chatting, inquiring about business, and taking an interest in the personal side. Like everything he does, these official private visits are not without reason. He makes these visits "so that he hears what he would not otherwise hear," explains Karola Block. And this is of some importance for those who want to forge alliances.

One of those who has worked in close collaboration with Schrempp over the years is Delene Ströh, who served as corporate relations manager in South Africa. She says she went through thick and thin with him. Jürgen Schrempp is not a simple individual, she says. "There are two sides to him. He is really two different people." On one hand, he is "an impressive character," she comments with approval. Yet, on the other hand, "He works hard and plays hard." As the first woman to make it to the top position in corporate relations, Ströh herself is not lacking in drive or self-confidence. Schrempp, known as a great communicator himself, cannot tolerate weaklings near him. So it is no accident that all those closest to him, from his years in South Africa to the present day, have strong characters.

Although he pushed her hard, on balance Karola Block's assessment of Schrempp is positive. She thinks of the time she spent working for him as "really great years," and she points out that he has the almost unique ability of sending people off "with a smile" even if he has just given them a thorough chewing-out. She puts this rather surprising feat down to his "great charisma." Schrempp is a polarizer, capable of dividing people into two camps. Waltraut Lenhard and Karola Block belong to that group of people who would do almost anything for Schrempp.

But even someone like Schrempp has to learn the hard way at some point. On one particular occasion after he gave Waltraut Lenhard a private talking-to, she was so angry that the next morning she went in and told him, "If you ever do that again, I will resign." Lost for an answer, Schrempp tried protesting his innocence: "I didn't mean you, I was talking to the people behind you." But both Lenhard and Schrempp were well aware that no one had been standing behind her. Lenhard did not earn the nickname "Sergeant Major" for nothing. She ran the office with military discipline and, if called for, she could handle the boss, too.

Johan Frederik van Olst admits that he was no particular friend of Jürgen Schrempp back then. They knew each other well enough, however, to meet regularly every Thursday evening and spend hours knocking a ball around a tennis court with other business colleagues, including Christoph Köpke.

For van Olst, who, like Schrempp, joined UCDD in 1974, this "nice, blond, blue-eyed German" did not prove to be a particularly good player, though he was blessed with the sprinting qualities of a Michael Chang. Van Olst has never forgotten Schrempp's style of play: "He is a fighter who battles for every point and doesn't give an inch."

Van Olst, now general agent for Mercedes cars in South Africa, says that Schrempp acted on the court as if he were playing for advancement in the company. Jürgen really did develop "the same personality and discipline as in his business life," and he played tennis "to win, not for the joy of the sport itself." Deliberately ambiguous, van Olst observes that "Schrempp loves the fight." In the process, Schrempp did not particularly care who his partner was. Rather, he "simply wanted to dominate—in tennis, in business, and in everything else."

It is precisely this attitude that saved the job of Martin Connolly. The Scot was working in Bloemfontein at the time, where he was responsible for the export of technical components. Dissatisfaction over Connolly's work led the then regional manager, Jacob Roos, to fire him—without having discussed it first with Schrempp, who had equal rank. Although he agreed with Roos about the need for his action, Schrempp apparently was annoyed that Roos had gone ahead without consulting him, and so he nullified the decision. Martin Connolly was reinstated shortly afterward. "Schrempp won six-love," says van Olst, continuing the tennis metaphor. "Never ignore Schrempp's position" was the lesson he drew from this incident.

Nobody negotiates as hard, as doggedly, and as consistently as Schrempp. He knows what he wants, and if necessary he will put in extra hours to achieve his aim. Theo Swart recalls when Schrempp was discussing the planned growth of the company, and the investment it would necessitate, until four o'clock in the morning—and after he won the battle, he happily went into the kitchen to make a batch of scrambled eggs.

Swart, himself considered one of the toughest negotiators in South African sales and now head of the McCarthy Group, South Africa's leading chain of car dealers (the chain sells one of every eight automobiles in the country), is all too familiar with Schrempp's doggedness, having more than once drawn the short straw against this stubborn German.

Schrempp's strategic farsightedness allows him to prepare in advance for such late-night sessions, but the real game is about power and money—and there is more to Schrempp than just his obstinacy or staying power. Swart has long been aware of how cleverly Schrempp acted at that time. The car dealer acknowledges that he too needed to come away from negotiations with the feeling of having come out on top, and at the conclusion of their dealings Schrempp gave him not just the impression that "in the end we both won" but "the feeling that I was important." In doing so, Schrempp got a significant ally on his side, for without a doubt Swart was one of the key figures, or even the only one, enabling Mercedes-Benz to conquer the South African market during the seventies.

During negotiations, says Swart, Schrempp goes to the heart of the matter "surely and directly." "And if you are not really well prepared," he adds, "then you will have a problem." It is a sure bet that Jürgen Schrempp is always very thoroughly prepared—and he is well aware, as almost no other could be, of his own position of strength.

My interview with Swart took place on the fifth floor of the McCarthy headquarters. From the balcony we had a beautiful view of Durban and the Indian Ocean behind it. Nonetheless, even if Jürgen Schrempp is a thousand miles away, he is invisibly present. "The quality of the product always allows him to take a hard stance," comments Theo Swart without rancor, despite the number of bruising encounters he has had with

Schrempp. Somehow this sounds like a belated declaration of surrender.

Schrempp is not a man for halfhearted compromises, as friend and foe alike must recognize. His former secretary Karola Block speaks from experience when she says that her boss "can be extremely hard." And Schrempp is not the type to try to solve problems on the phone. Instead, "he goes straight to the other person's office." This illustrates his clarity of purpose and unusual determination. "There is no middle way for Jürgen Schrempp—only black or white," says Theo Swart, and he should know. And the German can show emotion in the process—banging on the table and pacing up and down the office like a caged tiger. Nonetheless, Schrempp never leaves the room—either during the high or the low points of negotiations.

While Schrempp's reputation is that of an opponent to be taken seriously, he is not always a likable fellow. With a wry laugh, Hugh Murray characterizes the manager as "not a nice person."

Over the years, Murray has taken a professional interest in this shooting star and has come to the conclusion that Schrempp is "very, very ambitious, very, very hard, and very, very decisive." Because of their country's rich variety of fauna, South Africans tend to compare people with animals. For Murray there is no doubt about which group of animals to assign the Daimler man to: "Jürgen Schrempp is a predator with sharp teeth." From all evidence, it is not that this description is wrong, but it is by no means precise enough.

Kill or Be Killed

He would stop at nothing if someone stood in his way.
A Schrempp expert in South Africa

Jürgen Schrempp was working at his desk when he received the message that he should go up to see the chairman right away. Morris Shenker was not alone in his office—a renowned guest from the Group's headquarters in Stuttgart was with him. Gerhard Liener, Daimler-Benz's director of mergers and acquisitions (and hence also responsible for Daimler's associated companies) was on one of his five or six yearly visits to the branch office in South Africa.

The executives were faced with a problem unique in the company's hundred-year history: In 1980, following increased international pressure on the apartheid regime, the South African government felt that it had no alternative but to insist on the domestic production of engines. A licensed producer in Cape Town, Atlantis Diesel Engines (ADE), was given the mandate to manufacture diesel engines for all vehicle types. Like all the other automobile companies, UCDD (the forerunner of MBSA) had to decide whether to turn engine production over to ADE or build its own engine factory in South Africa. This presented Schrempp with the unique opportunity of being able to have input at the highest level.

The Mercedes man knew that the company's own development department was already working on a new V engine, but it was still considered too expensive to be a factor in the discussion about whether to relocate production capacity to the Cape.

When Schrempp entered the room he could not know that this face-to-face contact with Liener would turn out to be a decisive milestone in the further development of his career. He would always remember this first meeting with the influential board member responsible for mergers and acquisitions as "that famous conversation." When the experienced customer service manager was asked for his advice, he presented Shenker and Liener with his own idea for a solution to this problem. Schrempp's suggestion was that the six-cylinder V engine should be mass-produced—thousands at a time—on the assembly line. Upon his return, Liener approached the German specialists for their assessment of the viability of the project, and quickly gave his approval. "I took advantage of the opportunity to provide Gerhard Liener with all the information that was known worldwide about the current state of engine development," notes a visibly content Schrempp.

The director of mergers and acquisitions convinced his fellow board members at their next meeting of the project's prospects, and the board approved the plan. Mercedes thus participated in the engine tender, winning the order against tough American competition for the contract to produce engines for the South African vehicles. When, as a reward, Schrempp was offered a job in the mergers and acquisitions department, he turned it down with the appropriate expressions of gratitude. Schrempp calculated that having gained recognition in Germany for being a clever young manager was far more important to him than some boring desk job—and represented a success that would pay dividends in the future.

One should keep one's distance from elephants, rhinoceros, buffalo, leopards, and lions, as they are rightly considered the most dangerous creatures roaming South Africa's savanna.

The elephant's memory, the aggression of the rhinoceros, the Cape buffalo's irritability, the danger posed by a leopard, the confident power of the lion—if one imagines the human equivalents of these traits in Jürgen Schrempp, it becomes clear how he rose from being regional service manager in South Africa to the center of power at Daimler-Benz headquarters.

An excellent memory is one of Schrempp's outstanding characteristics. "All the facts are in his brain," reports Richard Wentzel, the company driver, adding that Schrempp can give speeches without notes. Like an elephant, too, Schrempp instantly recognizes people even if he has not seen them for many years. A decade after he had last seen Leo Borman, the former head of CDA, the two met by chance in the Namibian capital, Windhoek, where Schrempp was staying with other Mercedes heavyweights. Schrempp, by this time chairman of the board, "recognized me immediately, left Breitschwerdt and Gottschalk [other Mercedes executives] standing there, and greeted me," recalls Borman.

Anthony Church, a Mercedes dealer in South Africa, had a similar experience. "When Jürgen met me again at a board function in Möhringen," Church recalls, Schrempp left his fellow diners and "greeted me personally." Schrempp's reaction upon seeing Borman and Church is both characteristic and appealing. But it also shows that Schrempp has learned how to handle business associates in a way that keeps them on his side.

Is Schrempp ambitious? The answer is the same, no matter who you ask. "He is enormously ambitious," says Waltraut Lenhard, and she should know; after all, she worked very closely with him for many years. But could it be more than ambition? "Schrempp is an enormously ambitious person," confirms industrialist Brian McCarthy, "and he has a streak of ruthlessness." But any manager looking for success would need this; after all, the business world is not exactly known for giving the softest and sweetest little lambs a helping hand up the ladder. And what methods does Schrempp use to get ahead? Theo Swart, who knows Schrempp well, judges him benevolently. It is true that anyone standing in Schrempp's way is "swept aside," but it is done "peacefully, but firmly" and not discourteously, Swart says. But not everyone appraises Schrempp so amicably. If it suits his aims, he "stops at nothing," according to someone who has watched him closely for many years—and has good reason for wanting to remain anonymous.

Johan Frederik van Olst detects the typical characteristics of a lion in Schrempp.

The largest of the big cats lives in packs, and *Panthera leo* is considered a sociable animal. Nevertheless, the fellow occupants of a game reserve keep well clear of this predator, which cunningly and quickly breaks its victim's neck with one blow of its paw, or strangles it with the viselike grip of its jaws. Schrempp, too, is very strong, has a lot of influence, and is powerful and prominent. "He wakes up fighting," says van Olst, who was born in Cape Town; "he roars first" in conflict situations, and "his message is: I will kill you."

It is true that what happens in far-off Pretoria usually has few if any immediate consequences in Germany, and the methods a Mercedes manager uses to advance his career in a distant country are of no interest whatsoever to the German public.

So whether behind the outwardly pleasant, cosmopolitan façade Schrempp acts like a lion or, as one South African who has had close contact with him describes him, as a "hyena," few people take notice. But as this source explains, it is the hyenalike trait that, in combination with the others, accounts for Schrempp's truly unique career: "He knows when he can grab a mouthful of flesh, and when to keep his mouth shut."

Jürgen Schrempp left UCDD in September 1982 to become president of Daimler's subsidiary, Euclid Inc., based in Cleveland, Ohio, and devote himself to the task of restructuring the ailing company. He was forced to acknowledge, however, that Euclid, which produced heavy vehicles, was in such bad shape that it would be better to sell it off rather than attempt to resurrect it. It was a tough time, Schrempp says now of the year and a half he spent in Ohio: "Euclid was a hell of a job, and a real disaster." And even though he was not able to carry out his initial assignment and restructure the company, the way he orchestrated the sale impressed the then-chair of the Daimler board, Gerhard Prinz, and his soon-to-be successor, Werner Breitschwerdt.

When in April 1984 Schrempp returned to MBSA, he had already done some thinking about the direction he wanted his career to go in.

He knew he did not want to return to South Africa as a board member; he also knew that the current chairman and chief executive of MBSA, Morris Shenker, would soon retire from the post after twenty-five years of service. So Schrempp set his sights on being named vice chairman, from where he could be sure of succeeding Shenker.

Schrempp's opportunity arrived more quickly than he could ever have dared to hope, and in a manner that gave him no pleasure whatsoever. Morris Shenker, highly respected in South Africa and a comparatively restrained manager who could apply a strong hand and autocratic power when necessary, had to resign in February 1985 as chairman of MBSA for health reasons. But long before the announcement was made, Shenker and Liener, who were close friends with similar conservative outlooks and a long-standing business relationship, had long since decided in favor of promoting Schrempp. And Liener, who had for some time been Schrempp's mentor, had the influence required to make their decision a reality.

"He was the right guy. The toughest, meanest bastard," Hugh Murray says, quoting Gerhard Liener's appreciative opinion of Schrempp's abilities. Impressed by his drive and leadership qualities, Liener clearly endorsed Schrempp as Shenker's successor during his discussions with Daimler chairman Werner Breitschwerdt. And the vote of the director of mergers and acquisitions was considered decisive at the Stuttgart head-quarters.

On April 1, 1985, Schrempp became chief executive and chairman of Mercedes-Benz South Africa. When Shenker died of lung cancer seven months later, Schrempp acknowledged that he had lost "a close friend," and he says he will always remember that the dying Shenker asked him to be a pallbearer. "That was an unbelievably emotional situation," he comments, and it is clear that he was deeply moved.

Schrempp, who at this time was forty years old, found it difficult to follow in the footsteps of his famous predecessor. Shenker's life had been closely interconnected with the South African automotive industry. His successor faced enormous expectations at a time when economic and political conditions were extremely challenging.

Despite his behind-the-scenes maneuverings, Schrempp had cleverly kept a low profile in the years before and given Shenker room to move. "Shenker was the boss and Schrempp the employee" is how Theo Swart describes the pecking order. Now he was to be put to the test in a much more visible situation.

But Schrempp was not considered a brilliant manager for nothing. He had already successfully handled marketing problems, the tight sales market for Mercedes vehicles, and the unions. In the eight years before he left for his brief sojourn at Euclid in the United States, the German had already gained a reputation for his intricate knowledge of the whole process from manufacturing to sales and for being able to solve the problems involved at each stage.

Furthermore, Schrempp benefited from having the same basic atti-tudes, experiences, and traits as those that characterized his predecessor and had made him so strong. Like Shenker, Schrempp sought clear deci-sions and was ready to give his all for the company.

Schrempp was in every way Shenker's equal. But Schrempp had one advantage over the older man: While Morris Shenker's twenty-five years of activity on the board were distinguished by constant conflicts with his rivals, Schrempp knew the right way to go about dealing with such problems. His ability to win over important people to his way of thinking—and sweep aside those that stood in his way—was what took him to the top of MBSA. "He used his instinct when forming alliances," says a source who knows Schrempp well. Perhaps this is why the verdict on the methods Schrempp used to clear the way to his becoming head of MBSA is mixed. It was not just his business abilities that helped propel him to the top comparatively quickly, but also his character. "He was an opportunist during the whole of his South African career," comments the source. When it came to a choice between killing or being killed, Schrempp would accept only one option.

CHAPTER 2

A Lot of Flak

I am convinced that South Africa represents
one of the world's most promising investment opportunities.
Gerhard Liener

The very first Mercedes in South Africa, produced in Mannheim, Germany, and powered by a 1.5-horsepower engine, was introduced with great fanfare in 1897. Paul Kruger, the president of what was then called the South African Republic, even presented the vehicle's importer with a commemorative gold medal.

Gerhard Liener loves quoting that historic encounter with Kruger, the Boer president: "I believe that there is no other market in the world in which our launch was so notably celebrated." But what had begun so successfully was to be abruptly interrupted by two world wars. The first steps toward a relaunch onto the South African market were taken eight years after the end of the Second World War. In 1952 only a handful of automobiles were delivered; three decades later Mercedes-Benz was supplying fifteen thousand limousines and six thousand commercial vehicles annually.

Liener, the board member responsible for worldwide investments, is among the majority of the company's executives who judge Daimler-Benz's involvement primarily, or exclusively, in profit terms. When in 1982, years before the election of Nelson Mandela to the presidency of the Republic of South Africa in 1994, Liener was confronted by the question of increasing investment in the apartheid state, he praised the freedom of its market, its economic stability, and "its stable system of government." He uttered not a word about the repression, torture, and terror that was an everyday occurrence in South Africa at that time and made life for blacks, in particular, a living hell.

Instead, progress in the sectors of industrialization and the financial markets were what Liener cited as deserving of respect. "Daimler-Benz is one of the very first foreign investors" that recognized the opportunities offered by the South African market. Liener finished by acknowledging that South Africa has a modern economic system comparable with that of many medium-sized industrial nations in the Western world, even if there were some deficiencies that require improvement. By this he did not mean democratic development, but economic growth—in order to be able to continue to offer investors "excellent opportunities."

The "W class," as the Group's rich white customers were known, magnanimously praised the message from Liener and the company. Mercedes vehicles have been the number one status symbol of an almost exclusively white clientele for decades. That "has been a special problem for Mercedes-Benz," says Leo Borman, looking back.

Mercedes-Benz had about 45 percent of the market in luxury cars at the beginning of the eighties. When asked about the fact that whites' income was stagnating during this period, Morris Shenker, Schrempp's predecessor, said that this had almost no effect on sales of Mercedes-Benz vehicles, arguing that whites' incomes were stagnant only in relation to the income of the black population and not in absolute terms. He pointed out that his company sold as many as half of its cars to other businesses for use by their management staff. In addition, a considerable number were sold directly to businessmen, and a significant percentage to farmers. All in all, he said, he had "great faith in the income of those that use our vehicles."

When Schrempp took over the reins of MBSA, there was some speculation about whether Daimler-Benz intended to pull out of the country, like many companies faced with the difficult economic conditions prevailing at the time and the pressure to isolate South Africa economically in an attempt to force the government to dismantle the system of apartheid. In an interview with Hugh Murray of *Leadership* magazine shortly after Schrempp was appointed chairman of MBSA, the new head made it quite clear that he considered it imperative that they remain in the country. Mercedes-Benz "sees the great potential" of South Africa, Schrempp said at the time, and "as a result we are investing in this country." Even if business was not as good as it could be, the company had up till then "been extremely successful in every way . . . and not just from the point of view of profitability." He also noted that those who wanted to be successful in business needed to plan in terms of five or ten years.

These statements speak for themselves. But anyone who thinks that Schrempp's position was based solely on his efforts to increase market share and profitability have overlooked his second aim, which was the defeat of apartheid using all the means at a business executive's disposal.

Looked at from the moral point of view, Schrempp said, "continuing here" is the "right way to tackle it." And if economic conditions allowed it, he added, the company needed to "expand as much as possible." Schrempp boldly twisted the fundamental beliefs of those supporting sanctions to his advantage: "One must not overlook the fact that we are directly or indirectly responsible for the daily bread of 50,000 people." Sanctions "would perhaps harm precisely those people in whose name they were imposed."

The antiapartheid movement argued at the time that such notions were the reason that the world missed its chance to use sanctions to bring about the fall of the white-run government in 1985 or 1986. Convinced that he and his position were correct, Schrempp considers today that such accusations were unfounded and not particularly helpful: "If you have constructive criticisms to make, and if you want discussions with the authorities about their conditions for supporting reform and peaceful transition, then you must also be prepared to make your contribution." Withdrawal would have made mutual communication impossible, according to Schrempp. Foreign investment was a necessity if the country wanted to solve not just its economic "but, more importantly, its political problems."

At the time, however, South Africa was burning, and Daimler-Benz's involvement there had become apartheid opponents' number-one topic at shareholders' general meetings. Prior to Schrempp's arrival in South

Africa in 1974, Daimler-Benz had already been a partner in a project that delivered 137 vehicles for transporting South African army tanks, and Mercedes trucks were upgraded to army standards for transporting blacks to the "homelands" or for military operations within the "townships." But vociferous German antiapartheid activists pointed to more recent problems—conditions within the factories, the fact that the vehicles were sold virtually exclusively to whites, the delivery of Unimogs (transport vehicles) for military use, and the production of engines for the Ministry of Defense—as ways in which the Group was supporting the apartheid government.

They particularly opposed the way in which Mercedes got around German export regulations and sent thousands of dismantled Unimogs from Germany to South Africa by declaring that the Unimog parts were "machine components." It must be said, however, that there is a general problem worldwide with what are called "dual-use goods"—civilian products that also have military applications. ADE, a purely commercial company founded in South Africa, found itself in the same situation. They delivered engines for civilian use but also took part, as did other companies, in official government requests for bids. ADE's manager in Cape Town was not always able to keep track of which engines were eventually mounted in which vehicles. However, according to research carried out by the antiapartheid movement, tens of thousands of ADE engines were produced for armored vehicles and other heavy military vehicles.

Today Jürgen Schrempp will say two things about this form of support for the apartheid regime. He will point out that these allegations have all already been dealt with openly and comprehensively at Daimler shareholders' general meetings of past years. Furthermore, he will admit that the question of dual-use vehicles was "always a problem" for him. However, according to Schrempp, there will never be a satisfactory solution to the problem of dual use because companies can never be a hundred percent certain about the end use of their products—particularly when the political circumstances are so complicated.

This is as true for Daimler-Benz as it is for other companies that are active on international markets. From his point of view, there was no question that achieving the end of apartheid through reform was the better route, instead of bloody, radical change. In this regard, he says, despite all the individual criticisms, the involvement of Mercedes-Benz played a part "in fostering peaceful transition." Schrempp also stresses that in recent years many South Africans have come to see it in this light.

Angela Mai, one of the leading activists in the antiapartheid movement, is not one of those who see it this way. What she still finds "disgusting is that Jürgen E. Schrempp and all the others in positions of responsibility at Daimler-Benz coolly deny that they cooperated with the South African military." And she counters Schrempp's claim, saying, "Of course Mr. Schrempp knew that . . . all the Samil troop transporters were equipped with ADE engines." Even "every worker knew that the majority of the engines that they serviced had been delivered by the military." And that is something that "Mr. Schrempp, too, must have been aware of," says Mai, who is from Stuttgart.

"Why," she asks, "did Mr. Schrempp not take any action against it if he was really against supporting the apartheid regime?" She has a ready answer for her own question: "Of course he didn't do anything because the company knew right from the start what they had gotten involved in with ADE." Furthermore, "at the very least," Schrempp had "accepted that cooperation with the military was a prerequisite for the continued presence of Daimler-Benz in South Africa."

In addition to antiapartheid activists, German union leaders, including leaders of the powerful metalworkers' union IG Metall, demanded that Mercedes withdraw from South Africa. Concern began to grow among members of the Daimler-Benz board, especially about the effect of the continuing protests on the company's reputation.

Schrempp did not waver, however. He went on the offensive, advocating that the company remain in South Africa. He tried to win over the nine members of the board to his point of view, without glossing over any of the facts but using his instinctive tact. "Our leaving would be bad for the blacks," argued Schrempp, addressing the concerns of Daimler's chairman, Werner Breitschwerdt. "I was completely in favor of staying," confirms Breitschwerdt. There was also the question of whether or not the company's management should invest more millions in MBSA, the money-losing associated company.

Although Daimler-Benz had an assembly plant in Indonesia and factories in other foreign countries in the eighties, the situations there were not comparable with what was happening at the East London facility in South Africa, which the headquarters considered very important. Breitschwerdt says, "I considered Mr. Schrempp as important as our head at the Mercedes works in Brazil, but Africa was more difficult."

Schrempp's arguments ultimately prevailed, and the members of the board unanimously agreed that Daimler-Benz should continue to do business in South Africa. Schrempp, as chairman of MBSA, naturally had a significant degree of influence on board decisions. Breitschwerdt even considers that Schrempp played a vital role in this. "Mr. Schrempp was instrumental in Daimler-Benz remaining in South Africa," he says, and adds, "It is possible that we would have decided on a withdrawal if he had been able to justify one." But it did not come to that, in part or wholly because of Schrempp.

According to Schrempp, there is a direct connection between the involvement of companies, particularly Mercedes-Benz, and the "peaceful transition" in South Africa. In the mid-eighties, to support this process, Schrempp promoted a reform program that would involve black leaders in a constructive dialogue and define social responsibilities. The aim was to abolish apartheid by treating all employees in the factories equally, in adherence to the European Union code of behavior for companies operating in South Africa. Such demands may seem natural enough today, but at that time they represented an enormous challenge to the regime in Pretoria.

On the face of it, MBSA's chairman did not appear to be openly agitating. Schrempp makes a habit of avoiding direct confrontation; he seldom formulates remarks antagonistically. During public appearances he mostly disguises criticism behind a benevolent-seeming screen of economic considerations. For example, if he was speaking before representatives of the automotive industry with regard to the situation in South Africa, he would copiously emphasize MBSA's affirmation of South Africa as a business location. Then he would challenge the government "to accelerate their reform program." After this he would deal at length with rationalization measures in the South African engine industry, where he would make reference to the notion that only "the fittest will survive."

He would also make the point that every action undertaken had to conform to the law of the land, knowing that the law spoke the language of discrimination against alternative thinkers and blacks. On the other hand, he would say that "for us," meaning MBSA, there was "no discrimination"—that there were the same rights for all within the company regardless of their race, gender, or faith.

When directly questioned about his own attitude, the MBSA chairman at times would be more direct. In an interview with Hugh Murray, for example, during which the journalist asked him what he thought of apartheid, Schrempp referred to positive changes in the pass laws. He said he was for the abolition of apartheid "for humanitarian reasons," though he also said it was "impossible for any society to overlook restrictions that have such a large effect on the freedom of individuals to sell their work (and the fruits of it) in accordance with the laws of the marketplace." And this, he argued, was precisely what was "still tainted by apartheid."

Schrempp caught a fair amount of flak from high-level executives at the headquarters in Germany. The not-so-subtle message was "Take a good look at our company's customer structure," referring to the fact that nearly all Mercedes purchasers were white. One executive aggressively challenged him, saying, "If you were employed in the Soviet Union, would you also give speeches against Communism?" A more muted but equally unambiguous criticism was, "As guests in this country, we cannot make a political spectacle of ourselves."

While Schrempp's stance that Mercedes needed to stay in South Africa was welcomed, his personal criticisms of the apartheid regime, issued both within and outside the company, made him something of an outsider. Schrempp today refers to this with a hint of pride. When asked who his opponents were within the company, Schrempp avoids a direct reply. "I received clear and direct instructions from Stuttgart" is his deliberately vague response.

Werner Breitschwerdt is more open about the conflicts during that period, admitting that top managers at Daimler-Benz AG were "for the white government." And, unlike Schrempp, Breitschwerdt is also prepared to name names, citing Hans-Jürgen Hinrichs, who was the member of the nine-man board responsible for sales, and with whom Schrempp had to deal directly when he was vice chairman and then chairman of MBSA.

It is clear that Schrempp found himself in an extremely difficult situation. As contentious as some of the questions from German headquarters may have seemed, he was still able to energetically hold his own with factual arguments. But the fact remained that the company really was totally dependent on the custom of the white oppressors. It was only their high incomes that allowed Mercedes-Benz to sell so many of its luxury limousines.

Schrempp is proud that the condemnation from the board, instead of forcing him to abandon political responsibility, brought instead a new awareness of it. He says that from this episode he has learned that "a member of the board cannot disassociate himself from political reality." Ultimately the board member is "part of society and simultaneously an element that shapes society."

As fine as these words sound, the question remains: Why did such a large company not inform its executives about the real situation in South Africa sooner? Implausible as it may seem to Americans, the Germans never officially recognized the link between police repression in South Africa and the fact that they sold vehicles to this police force. Whatever the case, it is telling that Hans-Jürgen Hinrichs, director of sales and Schrempp's fiercest opponent on the matter of apartheid, was relieved of his duties only in November 1988, as a direct result of chairman Edzard Reuter's diversification policies.

Passive Protest from Pretoria

Mr. Schrempp's management was not holy, not holy.
Goodman Jordan, SAAWU unionist

A person's race, gender or faith are irrelevant
from the company's point of view.
Jürgen E. Schrempp, Chairman of Mercedes-Benz of South Africa

The South African Allied Workers Union (SAAWU) was the strongest union in the country's trade union movement at the beginning of the eighties. It considered its national action part of the international Communist revolution, according to Goodman Mlamli Jordan, an SAAWU member at the time. Their public statements were brutally suppressed by the white rulers, however, and free speech was nonexistent. Official legal recognition of SAAWU was expressly forbidden. Jordan believes that the ban on recognition was due to the fact that SAAWU, at least in previous years, had received financial contributions from one of the underground organizations that today make up the ruling party, the African National Congress (ANC).

In contrast, Les Kettledas, the national organizer for the National Automobile and Allied Workers Union (NAAWU), allowed the government to check NAAWU's financial records. This was intended to ensure that no money was received through Communist channels. In return for this, NAAWU was granted formal recognition. Hence SAAWU, with a larger membership, held a stronger position among the blacks, while NAAWU was the only negotiating partner acknowledged by the whites.

More than ten years have passed since Schrempp, as MBSA's chairman, was responsible for what went on at the East London works. But Goodman Mlamli Jordan still can talk in great detail about the former manager of the "opposing camp." As a member of various SAAWU delegations, Jordan met Schrempp twice during the mid-eighties, when the unions and MBSA held roundtable discussions on the strained situation at the factory in East London. It is ironic that my meeting with Goodman Mlamli Jordan, in the fall of 1997, took place in the Mercedes board's elegant meeting room in East London, where, years before, decisions directly affecting the lives of many employees had been made.

"Schrempp was a member of a board that provided structural support for apartheid," says Jordan, who still works for Mercedes in the East London plant. It is a condemnation less of MBSA's chairman than of his colleagues on the board. Schrempp's principles were clear to all, Jordan says; Schrempp signaled "that he was against apartheid," and his methods were "not heavy-handed." In the end the unionist sums the matter up nicely: "He had to take care of his business, so he could not appear hard."

Indeed, Schrempp was faced with a dilemma. In 1985 he characterized the situation at the Mercedes production plant as "very ragged." This reflected the political situation in the whole of the eastern province and not just in East London. It was very difficult for the management to cope with these circumstances. Schrempp, however, appeared optimistic about negotiating away the many problems with the workforce, as well as those between the management and the unions. There was unrest at the East London facility, with the predominantly black workers feeling powerless to respond to repressive acts by the predominantly white, right-wing foremen.

The fault line of the conflict split the factory, and black employees used "every opportunity" to stop work, including complaints about not having been given clean gloves or the toilets not having been cleaned. In the view of management, such acts constituted a "unilateral provocation."

On September 2, 1986, MBSA's entire black workforce went on strike. Employees demanded the reinstatement of thirteen of their colleagues who had been fired after they boycotted work for a day to commemorate a massacre by South African government security forces in Duncan Village, a black township near East London where many MBSA employees lived, a year earlier. The thirteen had also been accused by the company (which could not offer proof, according to NAAWU and SAAWU) of threatening coworkers who did not want to take part in the one-day stoppage.

"Those were the dark years of apartheid," says Les Kettledas, leaning back in his black leather armchair. "They were like war." The memory weighs heavily on Kettledas, who has had a successful career during the last few years under the Mandela government. In his office in Pretoria, South Africa's present deputy minister of labor has a staff of employees the size of which he could only dream of when he was leading negotiations against Mercedes.

Kettledas met Schrempp only on a single occasion. Nevertheless, he can remember the German clearly. "Jürgen Schrempp was the best chairman of the board in the German company's management. He was better than all the other top managers who had come from Germany." Schrempp had moved the company's philosophy forward, Kettledas says, and in the end simply didn't have "enough time for strong intervention."

Les Kettledas is not sure whether MBSA chairman Schrempp was actually aware of the real situation in East London. He notes that he and other union representatives never got to see the company's reports on the matter—which, he thinks, could have been part of a deliberate strategy. Also, the board of MBSA was in Pretoria, five hundred miles away from the center of unrest, and so it was the local management that carried out much of the political dirty work.

He has repeatedly asked himself what the former Mercedes chairman did about the reports he received from the management in East London. South Africa's deputy minister of labor considers it quite possible that Schrempp had not reflected sufficiently about the situation reports from the production works.

And, ultimately, in East London there was "no good management" present, says Kettledas, who thinks a moment and then amends his previous statement: The management in East London was "completely right-wing." Schrempp had let them react as they wanted, which allowed the "racist antiunionists" to put unpopular critics under enormous pressure or even fire them without further ado—without having to worry about direct intervention by the chairman. According to Les Kettledas, "there was a hard battle going on there between the unionists and the white racist management," and the rest of the company seemed to "forgive" the East London management for its actions.

Hundreds of workers were fired as retaliation for their political activities and demonstrations. A member of IG Metall, the large, well-known metalworkers' union in Germany, wrote that "this is nothing less than yet another impoverishment for the colleagues affected by these measures," and appealed to his colleagues at the Daimler-Benz plant in Mannheim, Germany, to "make use of your influence with the management and supervisory boards of Daimler-Benz to bring about the reinstatement of the colleagues at MBSA."

Schrempp's reaction to the problems in East London was by decree and through proper channels. "Solve the problem" was his internal instruction to the East London plant's management. Ultimately "he did not support management," says an insider who accuses the chairman of having "failed" to stand by the factory's management when they were in dire straits: "He simply increased the pressure they were under, instead of helping them." Perhaps, continues this source, "he didn't worry about the problem because he was not affected by it."

If Schrempp's verbal rejection of apartheid is not taken as definitive, then flaws begin to appear in the portrait that at first sight seems so unequivocal. Whatever the case, "the top management in Pretoria tolerated it all," says Les Kettledas, adding that Schrempp "could have intervened more" and "did not do enough to get rid of the white racists." Did Schrempp have the opportunity to do anything, or would such open conduct have cost him his job?

Change Despite Trade?

The positive example is called South Africa.
Jürgen E. Schrempp

Given his antiapartheid stance and the opposition of the Daimler-Benz board, Schrempp might well have made things easier for himself if he had voted for immediate disengagement. Or perhaps he would have made things easier for himself if he had cited only the economic argument for staying. But Schrempp, as he has done so often, chose neither of these simpler routes. He was firmly convinced that Mercedes needed to remain in South Africa if it wanted both to have political influence and—and here Schrempp was being a full-blooded manager—to earn a lot of money.

Certainly Schrempp's greatest success in the area of South African policy is the implementation of the European Union's code of conduct of 1977 (and its revised version of 1985) at MBSA's East London works. The declared aim of this EU code was to contribute to the "abolition of apartheid" in South Africa.

Schrempp himself is still proud of his intensive cooperation with unionists and particularly Franz Steinkühler, influential leader of IG Metall, the largest single union in the world. During the second half of the eighties, Steinkühler worked out a fourteen-point plan that would culminate in the complete equality of all employees at the East London production plant, and Schrempp gave it his support.

Under this plan, MBSA moved toward the "rejection of the currency of advantage offered by the apartheid laws." The right to strike was recognized, and union representatives in South Africa were guaranteed the same rights "as those enjoyed by representatives in Germany." Schrempp considers this internally implemented plan as an example of his "taking personal responsibility."

But even this success, so celebrated by Schrempp, has been criticized. Angela Mai, for one, does not trace the improvements at the East London Daimler-Benz plant back to Schrempp's performance alone, for as long "as there was no pressure being exerted, Daimler-Benz also stuck with the exploitative norm of all the other firms in South Africa regarding wages and social obligations toward the black workers." She believes that this "constant pressure from outside" was necessary. "Only then did the Mercedes management discover their social conscience and accept the inevitable." This is supported by "the fact that in 1986, while Mr. Schrempp was chairman of MBSA, wages 25 percent lower than those stipulated in the code were paid at a Mercedes subsidiary, MBEUS."

Of course, opinions still differ about the real influence of business on the peaceful transition that took place at the Cape. Brian McCarthy, of the McCarthy Group, considers it "important that Daimler-Benz remained in South Africa despite apartheid." The presence of Mercedes-Benz, as the largest automobile company, took on special significance after the Americans, British, and French had left the country.

Today, Schrempp cites South Africa as a prime example of the successful strategy of peaceful transition through commercial trade. He sees his arguments as receiving support from the most famous of all South Africans, who was imprisoned for twenty-six years and was later awarded the Nobel Peace Prize. "I am sure that Nelson Mandela, whom I greatly respect, also secretly sees it this way," Schrempp maintains. The high regard in which he is held in South Africa today may speak for the correctness of his contention. Mandela made him honorary consul general to South Africa in April 1995. And former German chancellor Helmut Kohl asked him to take on the chairmanship of German industry's Southern Africa Initiative (SAFRI), intended to improve the economic and political framework for cooperation between German business and the countries of southern Africa. Critics of Schrempp's attitude find it difficult to make their point when even *Leadership* magazine writes of Schrempp's reputation at the Cape: "As a friend and confidant of Nelson Mandela and Thabo Mbeki [recently elected president],

Schrempp is seen as South Africa's closest business ally in the world."

Supporters of the idea that Daimler-Benz should have withdrawn from South Africa during the struggle to end apartheid are not only to be found among the ranks of the antiapartheid movement, however. Leo Borman's argument that the change was brought about in response to United Nations sanctions and the "opposition from outside" is also diametrically opposed to Schrempp's theory. "In the end, South Africa was totally isolated," Borman points out, and suggests it was precisely this isolation that provided the decisive pressure. Despite this difference of opinion, however, Borman does count himself among Schrempp's numerous friends in South Africa.

In fairness, it must be pointed out that South Africa's isolation was not total—several German banks maintained a presence in the country throughout this period, along with MBSA—and that a peaceful transition took place there nevertheless. But it is also true that considering the desperate human rights situation during the mid-eighties, hearing South Africa being held up as a role model for "change through trade" must sound all too cynical for those who were on the front lines during that time.

In the end, it may never be clear whether Daimler-Benz's decision to continue doing business with the "W class" customers really helped bring about this fundamental change.

Profits Are Secondary

Without a doubt we are going through
our worst period at the moment.
Jürgen E. Schrempp, in the fall of 1985

During the years that Schrempp was chairman of MBSA, Daimler-Benz as a whole was riding an apparently endless wave of success. During the mid-eighties the Group's total sales exploded from $23 billion in 1984 to over $27 billion in 1985 and approximately $35 billion in 1986, not least because of the board's policy of diversification. Year after year the company posted considerable surpluses, ranging from slightly more than $532 million to slightly less than $1 billion. The number of employees worldwide grew from 199,000 to 319,000 with the creation of the new aviation and aerospace divisions—all reasons for the champagne corks to be popping at Daimler-Benz headquarters.

In South Africa, things were not so rosy. "We had very, very good years" and "we had made good profits" was the new chairman's summary of the situation Morris Shenker had left him, a few weeks after taking over the post.

But he also was aware—and spoke publicly of the fact—that there were "difficult times" ahead, both politically and economically. It was possible that the company would have to "inject money" in view of the country's long-term development potential. Schrempp's first year, in fact, was a financial fiasco that the company's management blamed primarily on disadvantageous circumstances: a dramatic drop in the exchange value of the South African rand (which hit the South African automotive industry, heavily dependent on imported parts, especially hard) and "continuing political unrest," as the Daimler-Benz chairman would later write in the company's annual report.

In the months after Schrempp took charge, MBSA improved the infrastructure at the East London works, and the Mercedes production plant was relatively peaceful—only after Schrempp had moved back to Germany and his successor, Sepp van Hüllen, had taken over would the situation deteriorate significantly. But all in all, and especially in contrast to the glowing success of Daimler-Benz as a whole, business in South Africa was bad under Schrempp—extremely bad. In 1986, Schrempp's second year in charge, sales melted away in the hot African sun and MBSA's boss had to report a dramatic 22 percent reduction in total sales, though in his annual report he tried to cast the situation in a better light, claiming that the outlook for orders and the number of employees were, "on the whole, satisfactory."

The numbers belie this attempt at optimism, however, for during Schrempp's second year, MBSA's operating deficit was nearly $22 million on sales of $4.6 billion. At the time, MBSA had the second worst balance sheet of all the companies with which Mercedes-Benz was associated, though it must be said that none of the others had to contend with such difficult political circumstances as did Schrempp's MBSA.

(Interestingly, despite the scathing remarks many make about Schrempp's successor, van Hüllen—for example, he fired five hundred workers who struck the East London plant in a wage dispute, and then promptly gave all of them their jobs back—MBSA's total sales rose by a good $100 million in the first year after Schrempp left, and MBSA slid into the black, having made the remarkable sum of $12.8 million. On the strength of this performance, van Hüllen was able to hire an additional 150 employees.)

Despite the challenges facing MBSA, it was only from spring 1985 to spring 1986 that Schrempp can be said to have gone all out to reinvigorate the company, meanwhile pinning his hopes on the market launch of new vehicle types, such as the W123, 200, 230E, 300D, and 280E. The decision to put these models on the market had been made three years before, however, and without his direct involvement, as he frankly admits.

Schrempp spent the second half of his tenure in South Africa preparing for his move to Germany.

Not everyone at the company is prepared to talk openly about the behavior of their chairman during this time—particularly when they are still financially dependent on the company. "One had the impression," says one employee, "that Mr. Schrempp had concentrated on the Schrempp career above all else," but adds that his image was better than his actual performance. Johan Frederik van Olst, general agent for Mercedes at the time, says he can understand Schrempp's "burning impatience for the next step," but criticizes him nonetheless for having "neglected" his duties at MBSA "because he was more interested in attending to his future." Olst says his former (and present) boss was "hardly still here" during the last nine months in South Africa.

The fact is that a large part of Schrempp's ability to ascend the Daimler-Benz ladder despite the negative economic performance in South Africa derives from the fact that the Daimler-Benz board, under the leadership of Werner Breitschwerdt, was more interested in Schrempp's integration skills than in his economic success or lack of it. Breitschwerdt says that attention was primarily focused not on the question of what MBSA could do to get back into the black in the short term, but on how the entire operation could be knocked back into shape.

In sum, the best that can be said of Schrempp's eleven years in South Africa is that they were a mixed success.

CHAPTER 3

The Company Way

I never found being chairman of the board attractive.
Werner Breitschwerdt, former Chairman of the Board
of Management of Daimler-Benz AG

Consider the stories of three Daimler-Benz chairmen. The first was overthrown by his successor after only four years; he could hardly bear the loss of power but abdicated in accordance with the rules of etiquette. He is still a welcome guest at the house of Daimler-Benz. The second was overthrown by his successor after eight years, could not bear the loss of power, and abdicated, but then took his revenge by writing a book about his experiences. He is not a welcome guest at the house of Daimler-Benz. The third is still in office and, thanks to his brilliant balance sheets, is a widely respected boss.

The first of these stories is that of Werner Breitschwerdt, now seventy. He clearly enjoys talking about the company that he loves above all, and he contentedly leans back in his armchair, lights up one of his small Davidoff cigars, and ruminates about how he came to be chairman. He had been working as director of development and, he says, made a "very good team" with his boss, Gerhard Prinz. When Prinz, who was one year younger, dropped dead while working out in his home gym, his distraught wife called Breitschwerdt, and Breitschwerdt's plan to continue his career tranquilly until retirement in the slipstream of Daimler's chairman came to a sudden end.

The post of chairman had never appealed to him, confesses Breitschwerdt, in contrast to his two successors, Edzard Reuter and Jürgen Schrempp, who had devoted themselves to this one objective. The greater responsibility became attractive only when the chairman of the supervisory board, Wilfried Guth, proposed it to him directly.

Guth, a heavyweight from Deutsche Bank, had a clear preference for Breitschwerdt over Reuter, an avowed Social Democrat who was the favorite of the metalworkers' union. After the choice of Breitschwerdt became known, Reuter engaged in a bit of mudslinging in the media, and although this initially frustrated Breitschwerdt, he says he eventually came to accept how much Reuter wanted the position of chairman, and later the two got to the point where they could relax with one another. However, Breitschwerdt's goodwill would be sorely tested upon the publication of Reuter's memoirs some years later.

Breitschwerdt's fall was partly the result of injudicious alliances. Instead of seeking a confederation with influential colleagues on the board, namely, Werner Niefer and Helmut Werner, he formed a coalition with members who wielded comparatively little influence. The real reason he lost his chairmanship, however, was his obstinate belief in traditional Daimler structures. Up to that time, the responsibilities of board members had been divided into development, production, and sales. But Reuter skillfully analyzed the shortcomings of this structure and, with Niefer's support, forced a restructuring along the lines of independent divisions such as vehicles or aviation and aerospace. There was more to this than just a concern for the company's future; Reuter knew about Breitschwerdt's conservative views on the structure of the company, and he positioned himself as a visionary prophet while characterizing Breitschwerdt's ability as limited to making small improvements in engine and gear systems. The acquisition of MTU (Motor and Turbine Union, producer of motors for war ships, tanks, and fighter planes), AEG (a manufacturer of industrial sensors), and Dornier (a producer of turboprop aircraft) allowed Reuter to complete this process of

presenting his boss as a pure technician who was not capable of direct-
ing Daimler-Benz's future; in the end, just about the only authority Breitschwerdt had left was that of signing contracts. Reuter had been
instrumental in the formation of the decisive structure and synergy
committee, whose membership excluded Breitschwerdt and whose
chairman, none other than Reuter himself, reported directly to the new
chairman of the supervisory board, Alfred Herrhausen.

But even this was not enough. Reuter needed the help of the media
to succeed in pushing through the early termination of Breitschwerdt's
contract. A never-ending stream of rumors about Breitschwerdt's
alleged incompetence was paraded through the political magazines over
the course of several months. Among the comments that were made
during this time were "[The] rather shy technician with few rhetorical
skills evidently [finds it] difficult to represent the firm publicly and
organize it internally" and, from the well-known *Der Spiegel,* "Right
from the start [Breitschwerdt] made serious tactical errors."

Herrhausen sealed the fate of the incumbent chairman of the board
by creating the new position of vice chairman and installing Reuter in it
in March 1987. As Breitschwerdt soberly sums it up, "the chairman of
the supervisory board and two members of the management board
were against me." Meanwhile, Reuter issued a declaration praising
Breitschwerdt that damned him with faint praise. "Nobody," he said,
"questioned [Breitschwerdt's] extraordinary quality as a development
engineer," and therein lay "his indisputable authority."

In a telling side comment, Breitschwerdt says, "When one is chairman of the board, one must occasionally just grin and bear it." He is
obviously unaware how accurately this sentence captures the nature of
Reuter's power grab and his own departure. At the supervisory board's
extraordinary meeting on July 22, 1987, Breitschwerdt was permitted to
"voluntarily" request the early annulment of his contract—one and a
half years before his obligation actually expired.

Edzard Reuter had achieved his aim. But the means by which
it happened would foretell the circumstances in which he himself
would later be brought down, and about which he would furiously
complain.

Hunting Elephants

Hyenas also go for elephant meat,
and an elephant's skin is damned hard.
A colleague of Jürgen E. Schrempp's

One of the factors that initially got Edzard Reuter a job with Daimler-Benz was likely his well-known father, who was at one time mayor of Berlin. Jürgen Schrempp's middle-class background, by contrast, conferred no advantage on him.

So it might have been a little surprising that when the two met a quarter of a century later in South Africa, both having gained management positions in the meantime, they found they got on rather well, developing a relationship that, at least during its best days, could be described as a close friendship.

Hugh Murray has known both men for many years, and it is his impression that Reuter, Schrempp's senior by sixteen years, "loved the young man like a son."

But the two differ in other ways besides background—in personality and in their views on management policy, among others. Reuter is considered a hands-off manager who delegates far too much and manages too little. No one doubts his visionary tendencies, but it is not for nothing that his competence as a manager is questioned. Schrempp's case is totally different: In South Africa he gained a reputation for being an energetic, decisive, involved, hands-on manager who kept his team on a tight rein.

His friendship with Edzard Reuter was one of the reasons, if not the principal one, that Schrempp's march to the top of the company was so successful. As financial director and, later, chairman of the board, Reuter assisted his protégé to the best of his ability: Reuter supported Schrempp's ascent to the Mercedes board, prompted Schrempp's shift to the chairmanship of Deutsche Aerospace, and initially chose Schrempp as his successor for the post of chairman of the board of Daimler-Benz AG. He then tried (but failed) to prevent Schrempp's election to chairman after he realized what was actually going on.

Reuter is considered the more intelligent of the two: "Reuter has forty percent more brain than Schrempp" is how a high Daimler functionary formulates it, though Schrempp makes up for this with "a mouth big enough for three." Furthermore, Schrempp has one ability that is largely lacking in Reuter: "He outwits people who are far and away more intelligent than he is," says one source, explaining the outcome of the power struggle.

The Fokker debacle, discussed at length later, is a good example of how Schrempp is able to master the kind of situation that broke Reuter. Schrempp was clearly the one who carried the bulk of the responsibility for the decision to acquire the company that would later prove so much of a burden, and so he elected to use the strategy of forward defense: He pointedly accepted all the blame, thus clearly signaling to the other members of the management and supervisory boards that they were also partly responsible but he would cover for them all. The strategy is disarming.

As aggressively as Schrempp had pleaded for the sale of Fokker, so he skillfully focused criticism of the diversification of aviation and aerospace, and the resulting catastrophic 1995 balance sheet, onto Reuter, who felt that Schrempp had betrayed him.

Reuter claimed that Schrempp had thrown every possible negative factor into the calculation of the loss. Further, he argued that the losses were in fact due to events at Deutsche Aerospace, for which Jürgen Schrempp had held sole management responsibility since its founding.

Despite the fact that Schrempp's radical restructuring of the "integrated technological concern" that Reuter had earlier brought into existence met with almost universal approval, his exaggeratedly negative presentation of the multimillion-dollar loss came in for heavy criticism from the supervisory board before his formal appointment as chairman at the shareholders' general meeting in 1995. More than a few people have spoken, off the record, about him "demolishing" his predecessor, and he is judged to have been "far more brutal than other managers."

"Our very existence would now be under threat if I had not carried out the restructuring," is how Schrempp justifies his approach. "We gave the company a new direction in just fifteen months. . . . In the end it is understandable that this was not an easy situation for my predecessor." Reuter, however, accused Schrempp of being interested only in "short-term profits instead of building for the future," which, when put so simply, was without doubt an unjustified charge. And, Schrempp says, as the relationship further deteriorated, "Mr. Reuter ceased to ask my advice after a particular point in time."

The story of the process by which Schrempp unseated Reuter is a long and complicated one. Its visible denouement began on June 29, 1994, when Edzard Reuter, who was still chairman of the board despite rumors—each more sensational than the last—that he was about to be replaced, delivered his report on the state of the company. "The first half year is almost over and we can report," announced Reuter smugly, "that the progress of almost all divisions is increasingly in accord with our declared published aims."

Things were going very well at Mercedes-Benz, he said. "Our car dealers in Germany continue to be up forty percent compared to the same period last year." And although he admitted that the German market was stagnant at best, Reuter went on to say with a huge smile on his face, "We can still speak with complete justification of the company's special economic situation." After all, "progress in some areas of business is going even better than expected."

Reuter ended his report with the conciliatory statement that work was being carried out "under high pressure in all divisions [to achieve] the very demanding targets." He pointed out that he was saying this "to expressly warn us all against assuming that we have reached a business-as-usual situation," and added that "there was still a long way to go."

Reuter believed at this point that he was a potential successor to Hilmar Kopper as chair of the supervisory board, despite the fact that in the previous few months it had been extremely difficult for him to do anything at all that met with Kopper's approval. While Jürgen Schrempp was poised to replace Reuter as the chairman of the board of management, Reuter was betting that Schrempp would pursue Reuter's dream of turning Daimler-Benz into a diversified automotive and aviation giant. So he merely smiled when Kopper rose and declared, "Only Mr. Reuter's readiness to shorten the duration of his chairmanship of the board of management has permitted today's announcement that the question of a successor has been settled." Kopper went on to say that continuing Reuter's tenure until the end of 1995 without making plans for a succession "would in my opinion have meant that the harmony so desired by all parties could only have been achieved with great difficulty." This statement echoed Reuter's own comment that "the disposition of personnel, particularly those managers who have been responsible for the commercial success," needed to be resolved soon, as well as his bold demand that "the speculation and uncertainties should be ended by clear decisions and, at least in this regard, harmony should be allowed to return."

Things apparently started out harmoniously enough. The motion to end Reuter's chairmanship on May 24 of the following year and simultaneously enthrone Schrempp received unanimous support. Reuter himself declared that he welcomed the early announcement of his successor. When Schrempp was told the result of the board's vote, he thanked the members of the supervisory board for the trust they had placed in him, he thanked Hilmar Kopper for his supportive words, he thanked his colleagues on the board of management, and above all he thanked Edzard Reuter for his support. But, as it turned out, the hope of harmony was soon to be shattered. Despite his and Schrempp's graceful-sounding comments, Reuter had no idea that by supporting Schrempp he had nominated as his successor a man who would put all his energy into realizing his own vision of creating the most profitable company in the world—a vision of a stripped-down company diametrically opposed to Reuter's concept of a far-reaching conglomerate.

Reuter would come to recognize his error months later, but by then it would be too late. He would be a spectator, helpless and ineffectual, watching his vision being taken apart decision by decision. In his despair, he would make one last attempt to save what was already past saving. But by the time he threw all his influence behind the idea that Manfred Gentz, a successful personnel director and the man responsible for Daimler-Benz Interservices (Debis), the flourishing Daimler division located in Reuter's hometown of Berlin, should become chairman of the board in Schrempp's place, the head of Dasa had already cultivated the support he needed to resist this coup attempt, and Kopper, the most powerful man on the supervisory board, had firmly decided in favor of Schrempp.

What Schrempp knew was that one does not become—and remain—

chairman of the largest European industrial undertaking by chance. The job goes to the individual who best promotes the interests of all the players in the game—industry, shareholders, customers, employees, and above all Germany's largest financial concern, Deutsche Bank.

Reuter clearly believed that Schrempp had something to do with diverting the support of Deutsche Bank chairman and board member Hilmar Kopper.

He said that while the banker had mentioned the "reservations of the shareholders," he nonetheless assured Reuter that he could be "sure of his vote" for the chairmanship of the supervisory board. Kopper, however, did an immediate about-face after holding a succession of discussions—"and intrigues, too, no doubt," according to an extremely frustrated Reuter. "All sorts of people were involved" in these intrigues, including those "who were pleased to call themselves my particularly close friends." Even though Reuter did not name Schrempp directly, it is clear whom he was aiming at.

Jürgen Schrempp was without doubt influential enough to have been able to convince his close confidant Hilmar Kopper to give Reuter the chairmanship of the supervisory board, especially as the banker was ready to do so. However, "I had my own opinion about that," says Schrempp, without wishing to add further detail to this comment. But all indications are that Schrempp was no longer prepared to stand by Reuter.

In Schrempp's long years at the company, many of his relationships altered significantly. As we have seen, the close father-son relationship with Reuter gave way to mutual recriminations. And another of his close relationships, that with Gerhard Liener, ended abruptly with Liener's suicide—the result of a series of events in which both Reuter and Schrempp played a part.

The Liener case raises explosive questions, the answers to some of which could be more than unpleasant for the company and any account of its history in the eighties and nineties. Many of them may remain unanswered, however, because those who were members of the select group of directors and board members at the time are keeping their counsel. Nevertheless, these events represent a painful chapter for Liener's friends and coworkers.

The matter began in late August 1994, two and a half months after the supervisory board's unanimous vote in favor of Schrempp as Reuter's successor, when Hilmar Kopper asked Liener, Daimler-Benz's financial director, to visit him at Deutsche Bank's headquarters. En route to the meeting in Frankfurt, Liener received the shock of his life when he opened up a copy of *Der Spiegel* and read this comment about himself: "The former expert on trucks cannot completely hide his lack of financial knowledge during meetings of the board of management. This is the price that must be paid as a result of the chairman of the board, Edzard Reuter, hanging on to the financial director for so long." The magazine's commentary concluded: "After a series of embarrassing performances, Daimler's financial director, Liener, must go."

Liener knew that five days earlier Reuter had invited Daimler's press chief, Matthias Kleinert, and an editor from *Der Spiegel* to his vacation cottage on Lake Constance, and he suspected that Reuter had planted the story during that session. But the fallout from the article was inescapable. At their meeting, Kopper informed Liener that he would have to resign at the shareholders' general meeting in 1995, two years before his contract would have expired.

"Of course, he didn't say that I had to go," Liener wrote later in a seventy-six-page document that was intended to be private but which was leaked to the press shortly before his death, "but expected that I would act in the interests of the company."

Liener was devastated. From his point of view, he was the one who had attempted to rectify the negative balance sheets Reuter had accumulated during his tenure as chairman, but here he was being cited as the cause of the company's financial problems. He confided to a close friend that he knew "we [Daimler-Benz] have suffered great losses" (as financial director, it was Liener's task to conceal the company's financial problems); this person approached Jürgen Schrempp, whom he had known for many years as well, and asked how he could let Reuter get away with this kind of character assassination. The new chairman of the board was reported to have answered, "I will deal with the matter as soon as an opportunity arises."

Indeed, the company's financial situation was subjected to a merciless internal examination, but by the time of the first shareholders' general meeting of the Schrempp era, in May 1995, the facts had not yet been made available; it took an additional six weeks to gather all the information needed. When the information was finally compiled, the situation was so bad that Schrempp was forced to issue a "profit warning" to a dismayed public and surprised shareholders.

In the meantime, the situation continued to deteriorate, and shortly thereafter Liener began writing a long document that detailed his impressions about how Reuter had systematically demolished him and cold-bloodedly gotten rid of him. Even in difficult years, Liener wrote, Reuter had been able to rely on him to do everything possible in his capacity as financial director "to conceal the sometimes catastrophic developments that were taking place as a result of [Reuter's] visions." And despite this, Gerhard Liener wrote, he was forced to come to the conclusion that "I was a nuisance and becoming increasingly troublesome."

When Liener penned this document, he apparently had no intention of making public his highly confidential analyses and accusations. The document was principally intended as a self-justification. For years he had felt that he had been under pressure from Edzard Reuter to present the financial situation in a more positive light than was actually the case. Liener was plagued by the worst possible pangs of conscience; he felt that he shared responsibility for this misrepresentation.

He was still debating with himself about what the correct reaction should be. He wondered whether he should write a book and publicize everything; evidently preparing this dossier helped him clarify the matter. Jürgen Schrempp is correct when he says that "putting the document together helped him find release." Liener's secondary consideration was to provide his closest confidants with an explanation of how he saw the mistaken decisions and deplorable state of affairs in the financial department—for which he held Edzard Reuter responsible.

In July 1995 portions of this document were published in *Manager* magazine, a German business publication, without Liener's knowledge or consent. The magazine's title for the fifteen-page cover story, "The Reuter File," went straight to the heart of the matter; the magazine called Liener's secret document the "fiercest possible reckoning-up with

Edzard Reuter and his era." But the article turned a harsh light not only on Reuter, of whom the magazine said, "Reuter is untruthful toward the shareholders, and would go down in history as the greatest destroyer of capital in postwar Germany," but on Hilmar Kopper, whom the magazine accused of a lack of concern about the $11 billion drop in Daimler-Benz's stock market value.

Liener learned about the publication of the dossier when he was on a trip to Mexico. The most appalling thoughts must have been going through Liener's mind during the flight back to Germany. Somebody must have abused Liener's trust; nevertheless, he knew that the breakup with the company that he so loved, brought about by the publication of this material, was inevitable and irreversible. In his mind, matters could not have been worse. The publication of what had been intended to be a private document had irreparably destroyed Gerhard Liener's life's work in a world where unwritten rules of discretion still apply.

Liener was picked up at the Frankfurt airport by one of Schrempp's closest friends, a colleague from the South African days who had traveled there at Schrempp's request to take their friend to Schrempp's home. "I'm finished," Liener told him on the way back.

Once face-to-face in Schrempp's house, the two old friends had the most difficult conversation of their lives because, after that article in *Manager*, Jürgen Schrempp knew that he had no alternative but to fire Liener immediately.

Several months later, on December 14, 1995, the sixty-three-year-old former financial director hanged himself at his house in Bad Wiessee.

Schrempp received news of the suicide during a visit to Cape Town, and he immediately flew back to Germany to attend the funeral of the man who had done more to further and shape his career in the company than any other. Schrempp accompanied Liener's wife, Margit, and the couple's two children to the graveside. A source who was there says that there were tears in Schrempp's eyes during the burial.

The company's management committee praised Liener's accomplishments in a statement issued after his death. From setting the agenda for globalization to launching Daimler-Benz shares onto the world's stock markets and particularly the New York Stock Exchange, "the deceased has always successfully made use of his far-reaching, predominantly international, experience in serving our company," the committee said.

Suicide is the act of a depressed individual, and it is certain that Liener had suffered from depression before, in the mid-1970s. Reuter said that when, as chairman of Daimler, he became aware of Liener's "almost incomprehensible business conduct," he ascribed it to "the effects of a serious illness," and so informed the head of the supervisory board, Hilmar Kopper. Kopper, however, perhaps because of the lack of any independent medical diagnosis of such a problem, "put off the unavoidable consequences, as he too saw it, for far too long," according to Reuter. Schrempp also agrees that there were psychological problems; Liener had previously been depressed, he notes, and he also points out that Liener's actions cannot be explained rationally.

Yet the trigger for his suicide seems to have been the publication of Liener's private dossier. That brings up the question of who leaked the document.

There are various estimates of the number of people who had obtained a copy of it before its publication, but it is clear that only a small group of insiders need to be considered. Schrempp puts forward the theory that Liener had distributed the manuscript "among twenty or thirty of his circle of friends," but this figure seems too high. And Liener would hardly have sent off the long text to so many people without giving some notice of its arrival through personal conversations hinting at his distress over the matter. Putting oneself in Liener's place, it would seem fairly clear that, in view of its explosive nature, he would surely only have given his secret document to his closest colleagues and friends. And he must have been aware that the chances of his trust being breached would also increase with each additional recipient.

Schrempp is quite clear on the matter: "He did not give me a copy of the document." And he points out that Liener did not do so "perhaps because of our discussion about the book." The financial director had earlier informed him that he wanted to write a book "about Reuter's conduct and the company." "I gave him some advice: Don't do it," says the chairman of the board, knowing the storm that would be stirred up with it. Schrempp says that when he told Liener of these reservations,

Liener was, for the moment, convinced. "You're right," Schrempp says Liener told him, giving in.

Another reason why Schrempp might not have received a copy of the dossier when so many others had may lie in the fact that a split between Liener and Schrempp had already occurred in 1994—even if Schrempp has, to this day, never said a word about what took place at that time.

Edzard Reuter claimed that Schrempp supported the demands for Liener's resignation and had "expressly confirmed that he too considered replacement long overdue" to both himself and Kopper. On the other hand, Schrempp did not want to confront his "personal friend" directly with this demand.

A South African colleague points out that "Schrempp is a difficult person" and "would hound you out of the company" if you dared stand in the way of his ideas. In the case of Gerhard Liener, the matter was perfectly clear, this source says: "He wanted Mr. Liener out quicker."

Schrempp himself vehemently denies this, not just because such comments damage his reputation but primarily because he does not see the matter in this light.

His memory is dominated by their painful parting, deep sadness over his loss, and the tragic death of a man who used to be one of his best friends. In Schrempp's view, he and Liener were never opponents. But this is not the view of others.

When Schrempp was appointed Reuter's successor as chairman and announced his new team for the board, Gerhard Liener's name was not on the list. Liener was surprisingly understanding about this decision by Schrempp, whom he considered "the most suitable successor" to Reuter. Schrempp would have to spend "ten or more years shaping a new epoch for the company," Liener announced with deceptive self-control, and he declared, "He has my support."

Why had Liener been so accepting of Schrempp's desire "to have a younger team around him when he starts" and to have Liener vacate his seat as financial director? Liener said at one point that nobody had helped him during "the difficult days," with the exception of "my friend Jürgen Schrempp." While Liener believed Kopper and Reuter had "thrown him off" the board, Schrempp created the post of personal advisor specially for him, which allowed him to remain in contact with his beloved company—and in style, too, in an office on the eleventh floor of the headquarters building, directly next door to Schrempp.

But then, at the end of July 1995, Schrempp terminated the three-year advisor's contract that he had with Liener. Why did he do so? The former financial director was to advise Schrempp on questions of capital markets and stock exchanges while on special international assignments. Liener's experience would have been of use to the company, particularly during the planned launch of Daimler shares onto the Shanghai stock exchange. And certainly the position gave Liener a feeling of security. While there was general sentiment among the board members that Liener needed to be ousted, should Schrempp have stood by his friend, retaining him as his advisor against what would certainly have been stiff opposition?

Gerhard Liener wrote at least two farewell letters, the contents of which have still not been made public, but the authorities closed their investigation long ago. The family has nothing to say on the matter, either, and their privacy should be respected.

Why, however, do nearly all the participants at Daimler-Benz maintain an almost complete silence? Why is not even one of them interested in exposing the individual or individuals who leaked Liener's dossier? Why do even journalists in Germany and South Africa remain tight-lipped on the subject while claiming that they know who passed the information on to *Manager* magazine? Is the man so irreplaceable that he must be protected?

PART 2

The Restructurer

CHAPTER 4

Not Exactly a Girls' Boarding School

You cannot afford to say no.
Koos Andriessen, Economics Minister of the Netherlands

In his spare time Schrempp plays chess with friends or with his sons Alex and Marc. He used to say that Gary Kasparov would be next in line after his children. Those who know Jürgen Schrempp are not surprised about his inner urge to defeat a world champion. This man is always pushing his limits, both privately and professionally.

The game Schrempp plays at Daimler-Benz is like chess in that it also is about making strategic decisions. Pieces are moved about, kings are crowned, and pawns are sacrificed as companies are bought up or cast off. Schrempp is a good chess player, but one particular gambit he made at Daimler-Benz nearly cost him his job as boss of the Group.

Deutsche Aerospace, with its Dornier turboprop aircraft and the large Airbus, already covered the lower and upper segments of the market. "Only the middle" was missing, in the opinion of Manfred Bischoff, now chairman of Deutsche Aerospace (Dasa). The gap—regional jets—was to be closed by acquiring Fokker, the Dutch aviation company. Fokker was the foremost supplier of jets with 80 to 130 seats, making it the ideal complement for Dasa's ascent to the top of the league of European aviation giants.

Schrempp and Bischoff, who was at the time Dasa's financial director, recognized that the deal presented both opportunities and risks. In theory the latter were considerable, but only if the worst-case scenario came to pass, with all factors suddenly becoming negative at the same time, from market slumps to price drops to unfavorable exchange rates. But those responsible at Dasa and on Daimler's board of management agreed that this confluence was well-nigh impossible.

During secret meetings, the negotiations had reached a decisive phase, and Schrempp was personally involved. Since the Dutch government had a 31 percent stake in Fokker, with a correspondingly high degree of influence over the company's business policies, Schrempp, Bischoff, and the Dasa team were dealing with Koos Andriessen, the Dutch minister of economics, and a number of other government officials. The talks were deadlocked, and Schrempp's tactics were deliberately provocative. On the other side, Andriessen was visibly annoyed with the German, who was repeatedly and relentlessly throwing up obstructions. After all, it is not every day that an economics minister has conditions for a deal of this size dictated to him by a company head.

The atmosphere was tense; all the participants were aware of the importance of the meetings. Fokker's significance in the Netherlands was comparable to that of Rolls-Royce in the United Kingdom, and so an attempt by a foreign company to acquire control was, as one of the negotiators put it, "as if we were to steal the crown jewels of Dutch identity."

Both Dasa managers were ready to buy, but only well below the price demanded by the government's team in The Hague. Furthermore, Schrempp was demanding an additional injection of about $188 million in capital, of which Dasa was prepared to provide half. The positions on both sides had hardened, and there was almost no sign of any readiness to compromise.

Koos Andriessen used every trick in the book. He stood up, went out, came back, and then stormed out of the room again in a fit of anger. In the end he stayed away for a whole hour, to hold talks with his own side (at least, that was the official explanation). "In reality," remembers Manfred Bischoff, it was all "tactical maneuvers." Evil tongues even say that Andriessen took a nap to soften up the waiting Dasa representatives. The mood, in any case, was explosive. The Dasa financial director, visibly angry, complained that the Dutch economics minister wanted to "drain us dry."

Schrempp would have liked nothing better than to indulge in a drink at this point, but during these negotiations alcoholic beverages were taboo, and Schrempp cursed the fruit juice that was the only thing available at the same time as he drank nearly a gallon of the stuff. And, in contrast to the extravagant catering that is usually a feature of corporate events, the Germans had to go out at four in the afternoon one day to pick up hot dogs at a stand nearby.

Perhaps the food and drink contributed to the Germans' irritation. "No, we're not going to do that," snapped Schrempp at one point, and he inconsiderately blew cigarette smoke toward his antagonists. Andriessen countered aggressively, believing that the Germans could not afford to turn down the Dutch offer. Nonetheless, Schrempp was fully aware of the strength of his own negotiating position. Dasa was developing the Euroliner in a joint venture with the French and British, which would have allowed them to enter the same market segment (though Schrempp kept quiet about the fact that the research and development for the Euroliner would have been in the $1.6 billion range, and it would have been much cheaper to purchase Fokker, whose aircraft were already flying).

In the end, the Germans prevailed, purchasing their desired stake in Fokker dirt cheap—only $283 million, which was practically the kind of thing Daimler could pay out of petty cash. Reuter, then the chairman of Daimler-Benz, expressed pleasure over the reasonable price (even though he feared that an injection of capital would be necessary) and forecast "an excellent initial strategic position" for the future, saying that the range of products Daimler's new partner, Fokker, would contribute would give the company "an attractive market position" in regional jets.

The negotiations were "completely successful," says Manfred Bischoff, "but not the developments that followed." That is putting it mildly. The takeover of the Dutch government's share in Fokker would soon turn out to be one of Schrempp's worst defeats.

Worse than Checkmate

I was sitting in my house in South Africa and had to cut the knot.
Jürgen E. Schrempp, *on the most serious decision of his professional life*

The most dramatic situation
[the Daimler-Benz Group] has ever faced.
Karl Feuerstein, at the extraordinary meeting of the
Supervisory Board in January 1996

"Jürgen, we can't keep Fokker," Manfred Bischoff, Dasa's financial director, told Schrempp bluntly in a meeting at Dasa's headquarters in Ottobrunn, Germany, a year and a half after Daimler's purchase of a share of Fokker. Bischoff was clearly upset. He was responsible for sales and profit, and the numbers he had on the sheets before him spoke of inevitable catastrophe. Schrempp did not answer. He had his heart set on the idea of creating Europe's largest aviation company, and what his friend was telling him ran counter to this dream. "We simply can't get away from the cost problem," Bischoff added, hoping for a reaction.

Schrempp maintained his silence, but Bischoff refused to drop the matter. "We have to keep plowing more and more into it," he said, and reminded Schrempp that they had already invested nearly $1.1 billion. "We can't go on like this."

Schrempp leaped out of his armchair and paced like a panther in a cage that was much too small. "Manfred, do you know what you are saying?" Both Schrempp and Bischoff were well aware of the consequences facing them: plant closures, job losses, thousands of layoffs—and, above all, the end of the dream of industrial supremacy in Europe. Bischoff felt he needed to push for an early decision, but Schrempp, atypically, asked for more time to think it through.

Five days before this portentous meeting, the chairman of the board of management and the head of Dasa posed the sixty-four-thousand-dollar question to the supervisory board in an internal memo: Should Daimler-Benz come up with the additional capital necessary to keep

Fokker going, or should the bridging loan, which had up to that time been extended by the Group itself, be discontinued? Fokker could continue to operate in its present form only if The Hague invested as much as Dasa. Since the Dutch government was not prepared to contribute a sufficient sum, the decision had, in effect, already been made for Bischoff and Schrempp, who by this time was now chairman of Daimler.

Things were going badly for the new head of Daimler. The Group's disastrous overall situation forced the supervisory board to hold a meeting that started unusually early and went on for six hours, during which the discussion was at times heated.

The company's civil aviation division was the subject on the table, and Schrempp's "love child," Fokker, (Schrempp always referred to Fokker as his "love child") was once again being subjected to fierce criticism. In his situation report Schrempp had left no room for doubt about the actual state of affairs. Fokker could not survive as things were, and the restructuring of the Dutch aviation company would require $1 billion or so in capital investment.

Deeply frustrated, Schrempp told the board that he could give no job guarantees in such a situation, and he added, "We have looked into every possible scenario regarding Fokker, from restructuring to filing for protection from creditors with a follow-up bankruptcy." The cost of bankruptcy proceedings could itself be as high as $1.3 billion to $1.6 billion.

Karl Feuerstein, as the leading representative of the workers union IG Metall and deputy chairman of the supervisory board, recalls that when the news spread, the crossfire of critical questions from employers and employees alike was fierce. He points out that Schrempp's "convincing presentations on the strategic advantages of the Fokker takeover" played a critical role in the decision to involve Daimler in the Dutch company, but he also says that the public sentiment at the time was that "buying Fokker was a corporate blunder."

Schrempp and Bischoff defended themselves as best they could. Bischoff cited "ruinous pricing by McDonnell Douglas" as one of the reasons for Fokker's financial troubles: "Deutsche Aerospace reacted too slowly [to McDonnell Douglas's actions]. And didn't react hard enough." Schrempp concurred, but even admitting that this was his fault was not enough to solve the problem.

Schrempp went to his farm in South Africa to think about his next move. In his teak-furnished library, surrounded by the classic detective stories and contemporary suspense books that he prefers, one of the mightiest men in the automotive industry ruminated over a decision that would have a direct influence on the lives of hundreds, if not thousands, of families in distant Europe.

———————

The January 1996 meeting of Daimler's controlling committee fell on financial director Manfred Gentz's fifty-fourth birthday, and although Hilmar Kopper made a show of wishing him a happy birthday at the beginning of the extraordinary meeting, neither Gentz nor the other members of the committee were feeling particularly festive. Following this, Karl Feuerstein, as speaker for the employees' representatives, reported that Daimler-Benz was facing its most difficult situation in its hundred-year history.

Schrempp took the floor after Feuerstein and came right to the point: Daimler had suffered a loss of over $3 billion in that financial year, the greatest loss ever reported by a German company. Every month during which Daimler continued to prop up Fokker cost another $94 million. Schrempp reminded the gathered committee members that they had done "far more" for Fokker than was usual "for strategically important companies" with which Daimler-Benz was associated. Nevertheless, he announced his decision: Dasa would have to withdraw its financial support of Fokker immediately.

Everybody in the room knew that this meant the end of Schrempp's dream of putting together the most powerful aviation group ever seen. Nevertheless, everybody understood the reasoning behind this decision, for the current figures and forecasts spoke for themselves. And everyone was also aware that they were a reflection of a misjudgment by Schrempp and Bischoff. They were the ones who had made the acquisition of Fokker so attractive to the management and supervisory boards. And it would be the most normal thing in the world for them to have to pay the price for this blunder.

In addition to facing the loss of his dream, Jürgen Schrempp also had to deal with the danger facing him at the moment. The only way he could save his neck was by immediately switching from the defensive to the offensive: "full moral responsibility" for the acquisition of Fokker.

The tactic was clever—very clever. Schrempp was giving his critics one opportunity to demand that the chairman of Daimler's board of management, Dasa's former chairman, take the only logical action. Those who allowed this offer to pass without taking advantage of it and demanding his immediate resignation would not dare try to force his ouster again at a later date. Schrempp was vulnerable, but he knew that if he survived, he would not be vulnerable again.

Manfred Bischoff was the first to react, and there was a whiff of pre-arrangement about it. He acknowledged to the committee that he had "collaborated closely [with Schrempp] in the takeover of Fokker" and that he felt "involved in the aforementioned moral responsibility." The head of Dasa made it quite clear that if the Fokker question was connected with Schrempp's fate, then "one would also have to discuss this matter with regard to my own personal capacity."

In his comments Bischoff, too, left no room for doubt about the Dutch subsidiary's disastrous position: A "further financial input in the region of billions [of Deutsche marks] is far more likely than a Fokker dividend payment." And Bischoff's conclusion was in full accord with Schrempp's: All further financial involvement should be brought to an immediate halt.

Karl Feuerstein, the employees' representative, nodded appreciatively from his place at the long table. He was impressed with the way Schrempp and Bischoff, sitting two seats to his right, had shouldered the responsibility for the Fokker acquisition. With Feuerstein's support, Schrempp realized that his plan had succeeded. By implicating not only himself but also all the other members of the supervisory board, via their decision not to call for his dismissal, he had expanded the circle of those responsible to include the members of the controlling committee. Snaring both employer's and employees' representatives in the net of earlier decisions created a sort of "we'll all hang together" mentality—and protected Schrempp from judgment.

Feuerstein conducted a formal ballot, and the result granted Daimler's chairman absolution. Jürgen Schrempp had played for high stakes and won the lot—and had discovered once again that he could rely on Karl Feuerstein.

But Schrempp's career was not the only one that had been put on the line at this meeting. The decision to withdraw from involvement with Fokker, together with the separation of Dornier's turboprop division and the upcoming dissolution of AEG, put the final nail in the coffin of Reuter's "integrated technological concern" and represented the definitive end of the Reuter era.

What Reuter could not have known as he left the meeting room that day was that this would be the last meeting in which he would participate as a representative of the employer's side.

Schrempp eventually did get his chess match with Gary Kasparov—and, as he openly confesses, it lasted all of two and a half minutes.

"In the end I couldn't move a single piece without suffering a deadly loss." The battle surrounding Fokker, on the other hand, lasted three years, but it too ended in disaster for Daimler's chairman.

For Jürgen Schrempp, the name Fokker will always be synonymous with his most shattering defeat and a disaster that remains unmatched in the history of both Dutch and German business.

Resignation Withdrawn

Schrempp: I cried because I couldn't save Fokker.
German newspaper headline

I never threatened to resign.
Jürgen E. Schrempp, on the Fokker case

In his capacity as chairman of the board of management of Daimler-Benz AG, Schrempp met with Dutch journalists the evening before the press conference on the company's annual report. The questions flew furiously: "Why did you buy Fokker? Was it really impossible to save

Fokker? Didn't you just want to buy up an unwanted competitor and get rid of it? In view of the money lost, are you considering resigning?"

Schrempp was fully aware that announcing a decision as important as Daimler-Benz's withdrawal from involvement with Fokker, a symbol of Dutch pride, solely on the basis of rational economic arguments would create a wave of disgust throughout the Netherlands. During his press conference Schrempp appeared visibly troubled as he announced that he had absolutely no choice whatsoever but to make the break with Fokker. Dutch newspapers subsequently reported that he was heartbroken, and that his wife could confirm that he had cried over Fokker—giving the impression of a sentimental restructurer.

"What is your attitude toward Germany, toward your homeland?" This was one of the questions from the throng of gathered journalists during this same press conference. Schrempp had knocked back at least five beers after the successful sales negotiations when Schrempp and Bischoff decided to sell Fokker in Frankfurt, and was in a good mood, according to someone who ought to know. He subsequently turned up at the Mercedes Museum, near Daimler-Benz's headquarters, a few hours later in a correspondingly relaxed frame of mind for the press conference.

"I'll tell you what I think of Germany. I don't want to have anything to do with Germany," Schrempp told them, visibly tipsy. Then, according to the German publication *Stern*, he added, "I'd rather watch impala." Many of the journalists there were utterly disgusted—as much by his behavior as by his comments. He "was bursting with self-confidence" was one of the kinder comments made afterward. Others spoke of arrogance, including Schrempp himself—referring to his feelings after selling Fokker, he replied: "Fantastic. You can call me arrogant if you want. I don't care."

The comment was published in the Dutch newspaper *Telegraaf*.

The question that inevitably arose of whether it was just the Fokker mission that had failed or whether it was also the failure of Fokker's chairman (and Schrempp's protégé) Ben van Schaik was a valid one. Schrempp, however, considered it absolutely unfounded. "On the contrary," the head of Daimler asserted vehemently. "As a Hollander in Holland," van Schaik had "carried out a balancing act [mediating] between public opinion, the works councils, and the government extremely well."

Schrempp had not forgotten that van Schaik, an expert in commercial vehicles, had immediately agreed to serve as Fokker's chairman when Schrempp asked for his help. And Schrempp was quick to express his gratitude: After his risky involvement with Fokker, van Schaik was not transferred, as a sort of disciplinary action, but was instead promoted to management chairman at Daimler-Benz of Brazil, a commercial vehicle manufacturer. "In Brazil I needed a successor for Eckrodt," who was leaving to become chairman of Adtrans Deutschland and vice chairman of Adtrans Worldwide, a joint venture between Daimler-Benz and ABB, which today is the world's largest supplier of railway systems.

When considered solely on the basis of the numbers, Schrempp failed at Dasa and van Schaik failed at Fokker. But what the German was able to achieve at the highest level within the company, van Schaik achieved one level lower. Both were climbing within the Group while the balance sheets were showing a downward trend. And while Schrempp consistently persecutes his enemies, he just as unfailingly rewards his followers. While Reuter's reputation lies in tatters today, Ben van Schaik's loyalty has paid off well for him.

As if Schrempp's comments at the press conference about wanting to go watch impala in South Africa were not sufficient to reveal his yearnings, the prominent placement of a small African statue below the South African flag in Schrempp's office would provide additional clues. By contrast, other souvenirs have their place on the low bookcase at the back of the room: the control stick of a Tornado fighter and the model of a Fokker 100 signed by the parties to the contract in the year when Fokker was originally acquired.

The occupant of the office on the eleventh floor of the Daimler-Benz tower makes decisions that reverberate well beyond Germany's borders. However, ruling from the heart of the Group can also be unbelievably lonely. The number of one's friends is inversely proportional to how high up one's office is. And the higher one rises, the steeper the fall.

"The whole world seemed to have the same opinion: buying Fokker was as easy as taking the train to the Netherlands," he reflects ironically. Schrempp still believes that "the decision to buy Fokker was strategically correct." To resign after the negative developments would have been "clearly the easiest way." But "it was my accursed duty to put things back in order." He is just like his old self when he speaks like this: a fighter, hard and self-confident. He is "really proud" that the employees also signaled their understanding and that, in the end, the divorce from Fokker was "clean, fair, and respectable."

He has been on the board of management of Daimler-Benz AG since 1987. And he has never once threatened to resign. "But I did consider this fundamental question on two occasions," the powerful manager admits.

One of those occasions was the conflict with Helmut Werner that came about as a result of restructuring the Group, to be discussed later. The other was the Fokker case.

When the question of responsibility is raised, Schrempp's attitude is this: "I have always said that it was my error," he declares in his forthright way. And he did offer to relinquish his office, as a member of Schrempp's trusted inner circle describes his stance at the time. Neither mentions that there was an element of calculation in this, as we have seen, in that this tactic undoubtedly enabled him to divert the supervisory board from any thought of demanding his resignation.

"When one has admitted making a mistake, there are two ways to react," explains Schrempp. "First, one goes. Or second, one corrects the mistake." He does not explain what "correction" might mean for those affected by the Fokker debacle.

In All Openness

The Industrial Sensors Division, thanks to its joint venture with
the Schneider Group, . . . now enjoys a position in the
world market that could not be better.
Edzard Reuter to the Supervisory Board, in March 1995

November 8, 1995: The members of the management and supervisory boards sat in silence in the meeting room of Daimler's headquarters while Ernst Stöckl, chairman of Daimler's subsidiary AEG and a friend from Schrempp's South Africa days, went through his sobering analysis of the current poor situation of all the subsidiary's divisions and projects.

There was no doubt that one of Germany's best-known companies, a major producer of industrial sensors, was heading toward collapse. But this all suited Schrempp's plan. His intention was to trim Daimler-Benz down, and what could be better than having Daimler's own board members, prompted by Stöckl's litany of horror, encourage him to pursue that goal? If anything went wrong, the responsibility would, again, be a shared one.

When Hilmar Kopper gave Schrempp the floor, shortly after nine in the morning, Daimler's chairman made his intentions abundantly clear. Loss-producing fields of business had no place within Daimler-Benz. The planned disinvestment at AEG meant that the structure of the division was no longer viable and, furthermore, no longer necessary. Schrempp announced the "quickest possible" wrap-up of unprofitable activities and outlined the few AEG projects that would, for the moment, remain active and would be directly assigned to Daimler.

The employees' representatives were conspicuously unhappy with this announcement. Karl Feuerstein tried to minimize, but could not completely prevent, the criticism being openly expressed by Bernard Wurl, a departmental manager and a member of the IG Metall union board whose opinion carried particular weight with the supervisory board. The otherwise conciliatory Wurl pointed out that there had been so much optimism about AEG's main trading sectors during the first half of the year, and he said that he found the "drastic change in the evaluation within so short a period" difficult to accept.

Wurl's censure was effective—not, however, against the present chairman but against his predecessor. During the supervisory board's discussion on AEG's prospects and its subdivisions, Edzard Reuter had predicted a future for the company that Jürgen Schrempp dismissed as patently absurd only eight months later.

Had the board been taken in by Edzard Reuter's promises? Had Reuter, having misjudged the realities, let himself become bewitched into making such a mistaken prognosis? Had he wanted to play down the difficulties involved in the diversification process that he so championed? Had the employees' representatives allowed themselves to be lulled into a false sense of security by the other board members? These are questions that can be answered only by those concerned—and they are unwilling to do so.

One can accuse Jürgen Schrempp of many things, but not of hesitance or restraint. Unwilling to gloss over AEG's situation or the consequences for its employees, he announced during discussions with employees about the sell-off of the industrial sensor firm that "it will be impossible to avoid making some clear adjustments in personnel." "In all honesty," Schrempp said, the "necessary improvement in the cost position would involve some loss of jobs." Schrempp's telling it like it is left a bitter aftertaste for the employees involved, however.

The amalgamation of the Frankfurt electronics group into Daimler-Benz AG was formally decided at the meeting of the supervisory board in April of the following year, thus hammering the decisive nail into AEG's coffin. The amalgamation would mean that AEG's unused loss carryover, amounting to $397 million, could be used by Daimler-Benz in the future.

While this was another element in Schrempp's plan to dismantle the "integrated technological concern" conceived under Reuter and restructure it according to his own lights, one thing should not be forgotten in any of the criticisms leveled at Reuter.

Schrempp was appointed acting member of the board of management of Daimler-Benz AG in September 1987 and became a full member in April 1989. From then on, Schrempp has been jointly responsible, along with the other board members, for all the decisions that had been made. Without exception.

When I ask him how he assesses himself and his restructuring methods for Daimler-Benz AG, the chairman of the board's reply is as honest as it is typical: "Nobody will ever spread a rumor about my having been brought up at a girls' boarding school." True. It was more a case of gunsmoke wafting over the scene than the sweet aroma of rose petals when Fokker was sold and AEG dismantled.

CHAPTER 5

King of the Snakes

He considers it his duty to help others share his success.
Advertisement for DaimlerChrysler AG featuring Jürgen Schrempp

Newspapers in Rome got some spectacular coverage out of a commotion that took place on Rome's famed Spanish Steps on Wednesday, July 19, 1995, late on that mild summer night. Shortly after midnight a young woman and two men of differing ages were strolling from the historic Piazza di Spagna in the direction of the Spanish Steps. The three were apparently relaxed after enjoying a small celebration, they were carrying an open bottle of red wine, and their voices echoed clearly through the night along the narrow alleys of Rome's city center.

Judging from their language, they were tourists from Germany.

Suddenly a police patrol appeared, inquired what they were doing there at that time of night, and asked to see their papers. The Germans were surprised at having been stopped and were not carrying any identification, according to the fifty-something man, having left their papers at the hotel. The woman was annoyed about being stopped and expressed this to the police in unmistakable terms.

Some newspaper reports went on to say that the two men intervened when the police wanted to take the woman with them to check her identity. There was a scuffle, and the pleasurable evening ended with all three being taken to a police station.

The three in question were Hartmut Schick, head of the planning team; Lydia Deininger, Jürgen Schrempp's secretary, who was celebrating her thirty-first birthday; and Schrempp himself.

The police report gave the details of the encounter, adding that the young woman had "used inopportune words toward the uniformed personnel," that Schrempp had gone back to their luxury hotel (which was only a hundred yards from the Spanish Steps) to retrieve their passports, and that after their personal information had been taken down, the three had been "allowed to leave without any delay."

"We had to go to the station with the police, where they wanted to write a report," says Schick, who finds it hard to understand what all the fuss was about in the first place. "We ordered a taxi afterward," he says, and summed it up this way: "There was nothing in the whole thing. We were not even on the Spanish Steps, just on the way back to the hotel."

Whether Lydia Deininger really called the two constables "stupid" is of secondary importance as was the claim that a policewoman had her hand injured during a scuffle and was given a medical certificate so that she could take four days off work. What is important here is that within just a few days, it became impossible to distinguish between fact and fiction in the press reports on "Schrempp's Roman nights" that flashed around the world. Headlines like "Head of Daimler Charged After Scuffle with Police" were common, and some articles claimed the complaints were "now being looked into by the justice department in Rome" and that "according to sources in the justice department, investigators have six months to decide whether to charge Schrempp and his companions."

One thing that can be established with certainty when trying to build up an objective picture of the events that took place in Rome is that some journalists and Schrempp opponents fell upon Daimler's boss like a pack of wolves, presumably in the hope of seriously damaging their victim's image or even using this to force him out of his job.

The public relations disaster could have been prevented if the company's press office had done even a halfway acceptable job. But the chairman of the board of management of Daimler-Benz AG was left to face the music by himself simply because that office was unmanned in July 1995. The highly savvy media person Detmar Grosse-Leege was suffering from serious health problems, Schrempp had promoted the experienced press officer Matthias Kleinert to be the Group's "foreign minister" that very month, and Kleinert's intended successor, Christoph Walther, was still obligated to work for the Hamburg cigarette producer Reemtsma until the fall of that year.

In his hour of need, the chairman of the board sought advice from the most senior representative of the Group in Rome. Schrempp met with Roland Klein, head of the press office there, and the president of the Italian branch of Mercedes-Benz for confidential discussions. He was given a promise of assistance on the principle that a mighty Mercedes boss can arrange anything in Italy. Schrempp, who at this point had been chairman of the board for just under two months and was still under the spell of the typical hierarchical way of thinking dominant at Daimler-Benz AG, agreed to let them take over.

Friends in South Africa were quick to spot the mistake. Hugh Murray, one of the best Schrempp experts on the scene, knows the strengths and weaknesses of his longtime friend. "Schrempp is the best communicator. When others take over the job for him," says the publisher of *Leadership* magazine, "it usually ends in disaster." The incident in Rome provides an excellent example of this.

As it turned out, the case was considered petty and neither Schrempp nor the others ever had to appear before a judge. But over the next few weeks his peers in upper management at Daimler-Benz "signaled Schrempp that this kind of thing was not permissible," says Schick. And "very few people called and stood by him at that time." So at the board's next meeting Schrempp raised the matter of the slip-up at the Spanish Steps himself, saying that it was all completely harmless and trying to play down the incident. Deininger, Schick, and he had only opened the bottle and were in no way drunk, and there had never been any scuffle. "The other members of the board listened to what he had to say and thought that there really was no cardinal sin involved. They reacted in an open manner," says a well-informed insider, assessing the backing given to Daimler's chairman from within the Group.

On the whole, it could have been a great deal worse for Schrempp and his companions. But the affair at the Spanish Steps has haunted Schrempp since the summer of 1995.

Schrempp has had to learn the hard way that he is no longer the lowly factory manager in distant South Africa where, far from any media interest, he could do or allow whatever he liked. Though nightly tours of the bars of South Africa were of no interest to the public there, much less in the rest of the world, a bottle of red wine is now enough to get Jürgen Schrempp onto the front pages of the international newspapers.

His problem is common to all those in the public eye. What is permitted to mere mortals is not at all suitable for the most powerful business leader in Europe. His private sphere is thus reduced to a minimum; journalists everywhere lie in wait hoping for the next Daimler disaster, and nothing would be more sensational than finding out that the boss is personally involved in it.

This is a threat of the highest order for someone of Schrempp's stamp. He is characterized by an agreeable streak of spontaneity the like of which is absent from all other top managers in Germany. Whereas other chairmen of large German companies are as stiff as pokers, Schrempp remains unpredictable and sometimes wonderfully refreshing.

"I take part in as few receptions as possible. I find them unpleasant on the whole," explains Schrempp nonchalantly. "I also stay away from some peer-group get-togethers"—an astonishing admission from the man some call "Germany's business chancellor." "Some have taken that badly," acknowledges Schrempp, but he maintains that such gatherings ultimately are "mostly just about being seen."

When the incident in Rome is raised in private, he still maintains that he acted correctly, though he says now that he "must recognize that there is not just the private person Jürgen Schrempp, but also the chairman of the board." And "one rightly expects" a chairman of the board to "act as an example." The incident at the Spanish Steps has left its mark on him. Speaking of his new status and the obligations it brings, Schrempp openly admits that "I do not always find it easy to appreciate the difference."

Indeed, Schrempp was too late in recognizing that this was a very serious matter. And he relied too much on the assistance of others. The chair of the supervisory board, Hilmar Kopper, whose relationship with Schrempp at that time could be characterized as "almost friendly," ordered the Daimler boss to his office in Frankfurt and told him in no uncertain terms that if Schrempp allowed himself the luxury of another such scandal, then he would be asked to leave—the banker's unmistakable threat to the manager, according to *Der Spiegel*. And Karl Feuerstein, too, made it absolutely clear that Schrempp's escapades needed to come to a halt once and for all, otherwise he would not be chairman of the board of management for much longer.

Even if Schrempp—with good reason—does not like to talk about it, the catastrophic media response and the pressure from the headquarters of the Deutsche Bank opened his eyes to the fact that he was living on a knife edge. As a result, he has thought things over, worked on himself, and learned to adapt to others' expectations.

The new Schrempp avoids the openness of restaurants; today he would sit on a hotel terrace in Rome. The new Schrempp would not do something like wander near the Spanish Steps around midnight in an inebriated state. The new Schrempp has also packed away his macho statements and keeps a greater distance than before when sharing an evening in the company of journalists.

"Things that didn't matter to him before," says Andreas Richter, chairman of Stuttgart's Chamber of Commerce and Industry, "do now." The new Schrempp is more mature and better behaved, more controlled, and therefore colder and less full of character. This development was predictable, though it is hardly positive.

Musical Chairs

Hilmar Kopper managed the restructuring
fairly and comprehensibly.
Jürgen E. Schrempp

The power struggle for the top post that had paralyzed the company's management for months at the end of 1996, keeping the media and the public on the edge of their seats and badly battering the company's reputation, was finally over. January 1, 1997, was the merger target date—the date when Mercedes-Benz AG became Daimler-Benz AG.

At the same time, the conflict between the head of Mercedes and the chairman of the board of Daimler was amicably resolved. "Mr. Werner believes that it would be impossible to make use of his industrial experience, as he has up to now been able to, within the framework of the new Group structure" was the public announcement by the chairman of the supervisory board after the meeting of the general committee on January 16. The Helmut Werner chapter was already history by the time of the extraordinary meeting held on January 23. The board's offer of new responsibilities at Daimler-Benz is "not in line with the demands that he makes on himself," Hilmar Kopper informed his colleagues on the supervisory board. None of them had any comment to make. Werner calmly left his post on the last day of the month—and at last pure sunshine reigned in the company.

It had all looked so different one year earlier. Schrempp's position in the summer of 1995 was as bad as it could be. The incident in Rome, the mistaken acquisition of Fokker, the layoffs associated with the "dollar-low rescue program" at Deutsche Aerospace, the upcoming dissolution of AEG—they had all contributed toward strong resistance to the Daimler boss, Schrempp, and given the head of Mercedes, Werner, a ray of hope. It came as somewhat of a surprise when Helmut Werner was given a second opportunity, after 1994, to become Lord of the Stars. And, as had been the case two years earlier, the Cologne-born Werner clearly had a better starting position than his opponent.

"It is of vital importance for Mr. Schrempp's survival that he succeeds in shooting down Mr. Werner and that he does not permit himself any further mistakes" was how one business journalist analyzed the situation. The analysis was precisely on target. If Schrempp wanted the top job, he had to prevent Werner from getting it. And he could not afford a second Rome.

But what strategy might allow Werner to become chairman of the board himself? How could Schrempp defend himself against the attacks of his rival?

Werner had just received a special honor—he was the only European to be included in a list of the twenty-five best managers worldwide published by the highly respected U.S. magazine *Business Week*. The apparently unconditionally supportive Mercedes men were not the only ones who thought Helmut Werner to be the incarnation of all that is good. And while Werner's image was becoming increasingly fanciful, Schrempp's stock—the hard-liner, the destroyer, the Rambo type—was sinking toward absolute zero.

Jürgen Schrempp would have been an exceptionally successful general. His analyses result in a clear picture of the situation, and any battles that he lost would be the result of the strength of the opposing forces. He would not lose on account of a poorly chosen strategy.

The rift between the Schrempp and Werner factions cut right through the Group and its committees. The Mercedes board members had largely closed ranks behind their boss while the Deutsche Aerospace group was just as supportive of their former chairman. The Daimler board of management was split. The supervisory board was also split, particularly as Feuerstein voted with Werner for the retention of Mercedes-Benz AG rather than folding it into Daimler-Benz. This was not a merger like DaimlerChrysler, but more like a restructuring.

Hilmar Kopper's support gave Schrempp not only the backing of the chairman of the supervisory board, but also that of the Deutsche Bank. And Schrempp appreciated what he had in his chief mentor: "He has stood by me through thick and thin," summed up Daimler's chairman. On one hand, "he doesn't interfere in my business matters"; on the other, "he is there when I need him," according to Schrempp, who stresses above all Kopper's fairness during the restructuring phase. In short: "He is an incredible gentleman."

Both opponents knew that there could be only one winner in this duel. As the final battle began, General Schrempp moved his artillery into position.

Schrempp's chance lay in the thorough restructuring of Daimler-Benz AG that had already been planned. In this regard the Daimler boss held all the aces. After the previous year's $3.5 billion deficit under Reuter, Schrempp introduced the reshaping of the Daimler-Benz portfolio and reduced the number of divisions by one-third, to twenty-three. The management of the divisions would no longer be in the hands of the individual chairmen of the Group's subsidiaries, but rather directly controlled by Daimler's chairman. Under Schrempp's scheme, the former Mercedes boss would, at best, have a role representing passenger car business on the new integrated board.

Schrempp had a chance of winning the debate on whether to retain the old structure or introduce a new one. But he needed to get his own battalions into position and simultaneously maneuver those of the enemy into a less favorable formation.

Werner's only hope was in creating and implementing a sensible structural alternative in which he could maintain his freedom as chairman of Mercedes. If this move failed, he would lose the autonomy he had enjoyed up until that time and would have to accept his fate, for the very existence of Mercedes-Benz AG would end with its integration into Daimler-Benz AG.

And thereafter only one person would make the decisions.

Schrempp's contract as chairman was scheduled to last until the turn of the century, as was that of Manfred Gentz as financial director. Hartmut Weule, responsible for research, had chosen to leave the company. The end of AEG was also the end of Ernst Stöckl's chairmanship. The two top positions were thus firmly in hand, and the new Group structure allowed for eight additional names.

Clearly, not all of the twenty-eight members of the board of management (made up of seven Daimler executives and twenty-one department heads) would survive the changes.

First of all, the six members of AEG's management committee were relieved of their duties. Even if the new integrated board was to be augmented by ten members, there would still be only five Dasa and four Daimler-Benz Interservices (Debis) department heads left. Furthermore, the nomination of other new members could not be ruled out. All those aiming for a post on the key committee, Daimler's board of management, knew that the selection process was really tough. Those who wanted to survive had to use their professional competence, management ability, and good conduct to earn the goodwill of the man at the top.

The main criterion, however, might have been unofficial but was the most decisive. There would be a crucial vote on the new Group structure at the forthcoming meeting of the board of management and the extraordinary supervisory board meeting in January 1997—and thus indirectly a personnel decision on the new chairman of the board. And Schrempp knew how to arrange who would take part in this vote and who would not.

Two proven Schrempp sympathizers already sat on the management committee of Daimler-Benz AG in the shape of the former Debis chairman, Klaus Mangold, and Dasa's chairman, Manfred Bischoff. Both owed a considerable part of their success to Schrempp and understood how they could repay him.

Schrempp had signaled years earlier that Mangold would be his first choice when a suitable opportunity for a post on Daimler's board came up, and in 1995 he ensured that his friend Klaus was nominated for the spot that had arisen.

"We have a personal, very good relationship," Bischoff openly admits. But Dasa's boss also had a significant interest in Schrempp's success. If Schrempp retained his position as front man, then Bischoff could maintain control of the chairmanship of the aviation and aerospace division for years to come. With Mangold and Bischoff, whose contracts ran until the year 2000, places three and four were also firmly allocated after those of Schrempp and Gentz.

In fifth place: Eckhard Cordes. "I am not a typical longtime Daimler employee" of Schrempp's, says the shooting star, who had in the meantime become the second most important man in the company. Cordes could still well remember his first meeting with Schrempp. As Cordes was preparing for his new duties as controller of Mercedes-Benz of Brazil in 1986, "a brisk man" came into Gerhard Liener's acquisitions and mergers office. They didn't exchange a single word, but Cordes got a feeling then that "he was something special."

It would be six years before they would meet again, not at work this time. Manfred Bischoff invited both of them to a private party. They spent a few relaxed hours together during a blind tasting of selected wines. "We had a lot of fun," recalls Cordes happily. First pleasure, then work.

The two first worked closely together on the Fokker purchase and then on the breakup of AEG. "We made up a small circle led by Jürgen Schrempp," a relaxed Cordes recounts easily, as if it had nothing to do with the end of a great, long-established German company and the elimination of tens of thousands of jobs. "There was no success in sight," says Cordes seriously, who also succinctly provides the reason: "We didn't have AEG under control." So the problem had to be solved somehow. Cordes disposed of the important parts of the Daimler subsidiary because Schrempp "had the guts to break it up," and he himself had then "managed the sale of AEG," impressing his boss in the process—a factor that was later to prove of great significance during the DaimlerChrysler deal.

Now the way was free for Cordes, who quotes the three ingredients he possessed to be a successful candidate for the board: "First, one must have behaved oneself properly. Second, many others must say that one had behaved oneself properly. And third, several positive circumstances must all come together."

This was exactly what happened in his case. He suggests, too, that there was "also a lot of luck" in his career. And it was easy to see how close these two at the top of Daimler were. When Cordes was asked to cite an example of a disagreement, however good-natured, with Schrempp, he demurred: "We mostly see the world similarly, so I couldn't really think of an example."

The question of why Schrempp created individual directorships on the board for the Motor and Turbin Union (MTU), Temic (part of Daimler-Benz that produces semiconductors, electronics for vehicles, and gas generators), or the rudimentary residue of AEG cannot be explained using the logic of the current Group structure. There was also a director of mergers and acquisitions; Cordes himself speaks of two departments that were in reality independent. Schrempp, though, placed his favorite at an important location. As director of mergers and acquisitions, he would be given control over the redirection of the Group's organization.

In order to hoist the economist onto the board of management, Schrempp canceled Cordes' nomination for the post of acting board member and promoted him to become a full member of this, the highest management committee, on April 3, 1996. Eckhard Cordes, for his part, discusses his directorship enigmatically and with humor: Group development combined with the direct management of industrial participation was "unusual" and "contrary to organizational theory," so it could be effective only "when the board member responsible got on well with the chairman."

In March 1996 the chairman of the board began dropping remarks about the vital necessity of restructuring the Group's holdings. He wisely refrained from mentioning the disempowering of the Mercedes boss. At the shareholders' general meeting of May 22, Daimler's chairman could bring a powerful argument to bear while giving his reasons for going ahead with his restructuring process: he could raise an operating profit of $1.3 billion from the depths of Reuter's deficit—which in the end amounted to $3.8 billion, according to calculations based on U.S. accounting norms. And although Werner's Mercedes-Benz AG contributed a profit of $1.5 billion and the aviation and aerospace division (Schrempp's former beat) accounted for the heaviest of the losses, the shareholders showed their gratitude with well-mannered applause—for Jürgen Schrempp.

Five seats were still free. The game of musical chairs entered its decisive phase, and at least one player was putting everything he could into winning.

The Battle of the Titans

Bernd Gottschalk is a powerful man at Mercedes-Benz. Or rather, he was. He led the commercial vehicles division for more than four years as Helmut Werner's successor and was thus one of the Group's leading figures. In July 1996 he became the first high-profile victim of the new Group structure that had replaced the old system of functional units one year earlier.

The more autonomy Spartan (the truck and transporter concern) and the U.S. subsidiary Freightliner were given, the greater Gottschalk's resentment. The commercial vehicles director had fought in vain against what he described as the "perforation" of his authority over the most senior of his subordinates. The setting up of a separate unit under Dieter Zetsche, who took over responsibility for sales in the passenger cars and commercial vehicles divisions in mid-1995, also involved a de facto reduction of the power of the commercial vehicles director. In the end Gottschalk himself could not make use of his declared good relationship with the employees' representatives, and so he left, as they say in good Daimler-speak, "at his own request."

New Group structures meant new positions, new power relationships, and new faces at the top of the Group. Places six through ten were occupied by the powerful Mercedes faction, the cohesion of which made it almost impossible to crack: Dieter Zetsche, Kurt J. Lauk, Heiner Tropitzsch, Jürgen Hubbert, and Helmut Werner. But now it was a matter of finding a way to break apart the closed society of the Mercedes men.

Young in years and rich in foreign placements, Dieter Zetsche was one of the relatively few top Mercedes managers who could contribute a rich fund of international experience. With the company for more than twenty years, Zetsche had gone beyond Europe's borders early on, starting his upper-level career in charge of the development department of Mercedes-Benz of Brazil; only two years later he moved on to become president of Mercedes-Benz Argentina, and after a further two years he became president of the Freightliner Corporation in Portland, Oregon. Such a career matched Schrempp's notion of international experience. And there is something else that should not be forgotten: Zetsche remained loyal to Schrempp, who rewarded him with the post of sales director on the board instead of a development department.

As a well-rounded executive, Dieter Zetsche was long considered the ideal candidate for passenger cars. Zetsche, after Hubbert the longest-serving member of the management committee for Mercedes passenger vehicles, had joint responsibility not just for successes, but also for failures such as for the problem caused by the Mercedes A-class not passing stability tests in Scandinavia. And whether he would survive this disaster as a member of the board was extremely questionable.

"Lauk will not survive in the Swabian Mercedes Mafia because he is an outsider," predicted one of the members of Mercedes' topmost team. The prominent manager's position on the executive floor was indeed a point of contention. That Kurt Lauk would establish himself at Daimler-Benz for any length of time was impossible. "No outsider has ever managed it," according to this manager's prognosis.

Nonetheless, there were two good arguments against this theory. Lauk, who had a doctorate in political science, was considered to be "highly intelligent." And, as vice chairman of the board of management of Audi AG, he had "picked a fight with Piëch [his boss at Audi—the chariman of Volkswagen today]," which was one of the attributes that first contributed to his positive image at Daimler. His career progressed correspondingly quickly: In August 1996 Lauk transferred directly from the board of VEBA (United Electricity and Mining AG) to the Mercedes board, took over the management of the commercial vehicle business, and began to be counted among the trusty followers of Mercedes-Benz chairman Helmut Werner.

Who would be crowned director of passenger cars? Helmut Werner, of course. Why Werner? An old adage has it that "only one king can rule a country." This is a principle that Jürgen Schrempp has consistently followed with great success.

They say that Helmut Werner really did meet with his colleagues on the Mercedes board and ask them one by one, "Are you behind me all the way? Will you support me against all those who are trying to meddle in our business affairs regarding this new Group structure?"

According to internal reports, they all gave him the nod. It is understandable that none of them can remember anything about it now. Zetsche, Lauk, Tropitzsch, and above all Werner's successor, Hubbert, would have to admit that at the time they did not have the guts to tell their boss the truth to his face. The corporate grapevine broadcast that they "lied bare-facedly," and it also referred to two meetings of an altogether different sort: Schrempp met separately with both Zetsche and Hubbert.

Yet again Schrempp demonstrated how clever he was. The Werner faction, which in the early fall of 1996 was still united behind their boss, was levered apart with the help of an enticing proposal. In a model of the personnel situation Schrempp presented to the board during the meeting of October 16, 1996, all of Werner's supporters would be represented on the new board. Gentz would give up his double function as financial and personnel director to concentrate solely on the financial department. Zetsche was given sales, Lauk the commercial vehicles division, and Hubbert passenger cars. Werner would be given a special status, and this was Schrempp's coup: The former Mercedes boss would be on the board and responsible for all vehicle business. This would leave only the posts of research director and personnel director vacant.

The inner structure of Zetsche's department appeared strange even to insiders. The question of why one of the ten board members had to be responsible for sales was difficult to answer. The point here, suspects a senior Dasa man, had less to do with the Group's new structure than with ensuring that the vehicle faction retained its power, and catering to the chairman of the board's own interests. Thus, in Dieter Zetsche, a Mercedes man was promoted into the top ten, and Schrempp could be assured of his gratitude in the future.

When it came to the new occupant of the research directorship, Schrempp skillfully played two completely opposite types against each other, only to give his support at the decisive moment to the candidate he could count upon to assist him when it came to the vote on the new Group structure: Klaus-Dieter Vöhringer, who in turn also expressed his gratitude for Schrempp's support.

Jürgen Hubbert had already quarreled with Edzard Reuter in years past. Reuter, with his authoritarian style, was accustomed to overriding any doubts that might have been expressed by the supervisory board. Even Johannes Semler, the shareholders' representative, was prompted to criticize Reuter's peremptory manner, and Manfred Göbels, the managers' representative, felt obliged to say that it was a "drama, a distinct case of self-righteousness" and that the former chairman of the board's approach was "truly awful."

Reuter repeatedly proclaimed that he wanted no cringers on the boards. "In reality there was no open corporate culture," recalls Göbels. Jürgen Hubbert was one of Reuter's victims. When the Mercedes man dared to ask critical questions at a supervisory board meeting, Reuter attacked him harshly: "Nobody is forcing you to participate here." Since then a state of open war had existed between the two. Jürgen Hubbert was interested in a fundamental revolution.

It would appear to be wrong to accuse Jürgen Hubbert of harboring ambitions of destroying Werner so that he would gain power over the passenger vehicles division. Hubbert held an important post under Werner, as he would in Schrempp's planned structure: head of Mercedes' passenger car business. That gave him no reason to intervene against Werner, even if the relationship between the two managers was considered to be extremely tense.

Heiner Tropitzsch joined Daimler as a twenty-seven-year-old in 1969. Twenty years later he assumed responsibility for the personnel department on the Mercedes board of management and was, for the moment, loyal to the Mercedes-Benz chairman, Helmut Werner. But Schrempp's gambit of taking the personnel department from Gentz opened up an opportunity to elevate another Mercedes man to Daimler's board of management: Tropitzsch. And as a result, it was now also clear whom Tropitzsch would support in a borderline decision.

The day of the crucial board meeting was Helmut Werner's sixtieth birthday, a good reason for a celebration with his best friends. Schrempp congratulated him during a very personal speech in which he announced his "closeness and friendship" with the man from Cologne. In contrast to the widely held belief that there was a personal battle between the two of them, Werner repeatedly insisted that the clash was based on conflicting ideas and in no way the result of personal animosity. The Mercedes man did admit, however, to "a certain enthusiasm for constructive conflict," which, according to Werner, was something "we both experience most intensively." Nicely formulated, Mr. Werner.

His plans in place, the board meeting developed into a total triumph for Schrempp. Daimler's board members voted seven to one in favor of the new Group structure with the transfer of Mercedes-Benz AG to Daimler-Benz AG that this would bring about. Just a few weeks earlier nobody would have believed that this resolution would have been accepted by such a wide margin. Schrempp had brought off the difficult trick of splitting the successful Mercedes team—which up to this point had contributed more than 70 percent of the Group's total sales.

In December 1996 Schrempp and Gentz applied to the supervisory board for a vote on the restructuring of the Group at their forthcoming extraordinary meeting on January 23, 1997.

On January 17 Hilmar Kopper informed the members of the supervisory board, in advance and confidentially, that the previous day the general committee had looked at point three on the agenda for the extraordinary meeting of the supervisory board, to be held in a week's time, and "recommended" five new members of the board of management "for approval."

The recommendation of Jürgen Hubbert, Kurt Lauk, Heiner Tropitzsch, Klaus-Dieter Vöhringer (as head of research), and Dieter Zetsche meant that at a stroke five Mercedes board members were promoted into the top ten of Daimler's management. Their terms of office would all run until March 31, 2002. One name was no longer included in the new Group structure, and Helmut Werner knew that the battle was lost.

The rival was vanquished, the chairman of the board untouchable for some years to come—and still victory was not complete. The second phase of Schrempp's battle plan now came into effect: It was time to make peace with the man his brilliant strategy had enabled him to knock out of the race. Hence the attractive offers (vice chairman of Daimler's board of management, for example) made to Helmut Werner really were intended to be taken seriously. Werner's threatened departure was being treated with considerable concern. "We are dreadfully afraid that the value of our shares will fall if he leaves," revealed one of the executive team during a confidential conversation.

But Helmut Werner had decided—against Daimler-Benz and, above all, against Jürgen Schrempp. He had been too deeply wounded by his former friend: first the lost battle for the chairmanship of the board of management during the summer of 1994, and then the curtailment of the power he enjoyed as the unconstrained head of Mercedes-Benz.

"The coordination of the entire vehicle sector must be under the control of the chairman of the [Mercedes] board of management," the Mercedes boss demanded before the decision. But as he was unable to get his own ideas past either the board of management or the supervisory board, lasting conflict with Daimler's chairman was all that could be expected. "We must not preprogram any permanent cause of friction with chairman of the board Schrempp," declared Werner, who consequently took his leave. Many still find it hard to understand why he left. Those who study Schrempp's personality more closely know the answer: Only one can give the orders.

One week before the extraordinary meeting of the supervisory board, a visibly unnerved Werner had already thrown in the towel, and shortly thereafter he formally resigned from the boards of both corporations. The apparently impossible had happened.

At the meeting on January 23, the comprehensive situation analysis that is usually given at such meetings was dropped in this case, and Schrempp gave the supervisory board's members only a short status report. And even the new Superjumbo A3XX—which would be the largest airplane in the world if the decision was made to build it—elicited only a single banal inquiry by one of those present.

Schrempp began his discussion of Daimler-Benz's future with a report on twenty worldwide companies. This in-depth comparison impressed even the supervisory board's members, despite the fact that the chairman of the board of management had consistently kept them up to date on the progress of his research. Schrempp also made it clear that "we have drawn on expert opinions from the world's leading business schools and universities." In the end "several model structures" for Daimler-Benz AG were suggested that were then weighed against each other by the top management.

All those present were aware of the significance of this meeting. Schrempp had happily listed the numerous weaknesses of the management concept that had been in place up until that time, and he stressed the advantages of his new Group structure—which would, incidentally, also dispose of the question of who would lead the company into the new millennium. "We must devote ourselves to our actual areas of interest," Schrempp demanded of the assembled members of the management and supervisory boards. And then they went on to the vote. But Schrempp already knew he could rely on the broad support of the supervisory board, in view of the prior endorsements provided by the board of management and the general committee.

When Schrempp's wealth of ideas was compared with the lack of any definite proposals by his opponent, it was hardly surprising that Werner lost the battle for first place in the company. Werner wanted to retain the autonomy of the automobile manufacturer and tried to defeat Schrempp's new structure without presenting any alternative of his own. The discrepancy between what he needed to do to win over his fellow board members and what he actually did was spectacular.

The victory was ensured by the fact that Werner had barricaded himself in his supposedly safe Mercedes fortress, while Schrempp had aroused the power interests of the Mercedes board of management. In the end, not even the mighty Mercedes board stood by Werner; the automobile faction's concerted action failed to materialize. In the clash between those who were for structural change and the one who was against it, the Mercedes chairman was fighting a battle whose outcome had already been decided.

The irony could not have been greater: He was the only German selected by the international press to be included in the list of the world's best managers. Shortly thereafter a frustrated Helmut Werner announced that he would leave the company. There was considerable consternation. Daimler-Benz without the charismatic Mercedes boss was hardly imaginable. And the situation created an image problem: In public, Helmut Werner had to be praised, while simultaneously his reputation within the company had to be tarnished in order to limit the fallout of the shift. As part of that strategy, a number of questions were immediately raised within the company. Was it disadvantageous for the organization when one man, who had not always had a lucky touch at Mercedes, left? Why had Mercedes posted a loss of $957 million in 1993 under Werner's leadership? Was Werner's lack of sufficient courage to make the necessary innovative decisions behind the fact that the truck sector had gone into the red in Europe? And why hadn't the C-class achieved its sales targets?

Suddenly, the loss of Helmut Werner seemed bearable. "He does not have the charisma of a Jürgen Schrempp" was the new line on the top floor of Daimler-Benz AG. This sentence would never have passed the lips of any of Daimler's top team had Werner accepted the post of vice chairman of the board of management, but it is a fact that allegiances can change that quickly at the Group's headquarters.

In contrast to what was being bruited about within the company, other words were selected for Werner at his last supervisory board meeting. Kopper's approbation knew no bounds: What had Werner not done for the company, the employees, and the management team? Hymns were practically chanted for the departing Mercedes manager. He had served "in special ways," he had contributed to success "with his exemplary efforts"—the litany of praise could be continued ad infinitum.

Observing the proprieties, Werner thanked Kopper for the warm words. He had reached his decision on the basis of "inner convictions," declared the Mercedes boss in a personal statement. He regretted that much had been inflated out of all proportion by the media. Then he raised the white flag and agreed to the declaration of capitulation that had been laid out for him. Without making any fuss, Helmut Werner gave his approval for the assets of Mercedes-Benz AG to be transferred to Daimler-Benz AG "in relation *inter se* with effect from midnight 31 December 1996 / 1 January 1997," according to the wording of the amalgamation contract.

At the next shareholders' general meeting, Hilmar Kopper declared that "Mr. Werner has left the board of management at his own request, before the expiration of his contract." All's well that ends well for Jürgen E. Schrempp, the henceforth undisputed Lord of the Stars.

So Helmut Werner followed Edzard Reuter in losing a power struggle against Schrempp. But in contrast to Reuter, Werner obeyed the rules of etiquette—and this was handsomely rewarded with a seven-figure settlement. The only condition? That he maintain his silence.

Many more people would be prepared to maintain their silence for the sum of $3,006,550. Even Daimler boss Schrempp would have to work for two years to earn what Werner received for his departure from the boards of Daimler-Benz AG and Mercedes-Benz AG. At least in the future nobody would have to worry about the multimillionaire's financial well-being.

When the supervisory board approved Schrempp as Reuter's successor in June 1994, he was aware of the problems that his predecessor had left behind. It can be safely assumed, however, that he did not guess their full extent. "I have never before seen a company with such low morale, such terrible communication, and such bad blood" was Hugh Murray's judgment on Daimler-Benz at the close of the Reuter era.

The South African could still recall a private conversation in which he told his friend Schrempp how another expert on the Group described the situation during the time of upheaval: "Möhringen was a snake pit" was the disparaging verdict on conditions at Daimler's headquarters before Schrempp took office. When Schrempp heard this comparison, he clapped the publisher of *Leadership* magazine on the shoulder and retorted with good humor, "And I am the king of the snakes."

Schrempp's achievement becomes clearer when one looks back to this phase, when Daimler-Benz was facing a critical economic situation and Schrempp took his place at the center of a company that had gotten out of step. No one else could have succeeded in getting the powerful vehicle faction to break ranks and vote against Mercedes chairman Werner, in bringing the Daimler board into line, and in persuading the whole supervisory board to support the new Group and personnel structure. And he managed to convert the decrepit dinghy called "Bullshit Castle," as Schrempp always used to describe the Group's headquarters, into the highly modern flagship of German business, which by steaming full speed ahead has been able to break one record after another.

Schrempp's PR advisor Detmar Grosse-Leege was quite right when he pointed out that "the so-called battle of the titans only ever [existed] in the magazines, and in journalists' heads." It was indeed true that one of the competitors hardly fought at all and the other gained a famous victory easily and without any great effort. Bravo, Mr. Schrempp!

CHAPTER 6

Red Balance Sheet, Red Carpet

Schrempp was clever enough to simply give up.
A high-ranking Deutsche Aerospace official

It was July 1986, and the Group's corporate logic spoke a clear language: Engineer Johann Schäffler must be promoted to become the first boss of the newly created Deutsche Aerospace AG. As chairman of the business management team of Messerschmitt-Bölkow-Blohm GmbH (MBB), the largest manufacturer of military aircraft and helicopters in Germany, he brought with him many years' experience and all the qualities necessary for this position. Schäffler was rightly considered to be *the* German aviation expert and was therefore given responsibility for the Dornier Group's aviation and aerospace division on Daimler's board of management. Schäffler also had the best connections to a particularly important group—high-level German politicians.

In Germany, both civil and military aviation depend, to a greater extent than almost any other branch of industry, on the flow of subsidies from Germany's Ministry of Defense and the ministry responsible for research. "[Schäffler] has great standing in Bonn," confirms one long-term Dasa employee, for whom no questions remained about who would become Dasa's first chairman.

In July 1986 MBSA chairman Schrempp was preparing for his

departure from South Africa and his return to where his international career had begun fourteen years earlier. The commercial vehicles director, Gerhard Liener, had attracted Schrempp to the Group's headquarters using all the means at his disposal, and had pushed through Schrempp's nomination to Daimler's management committee. Liener appreciated his knowledge of trucks, as well as the bond of their deep friendship, which was why he had arranged for him to be department head responsible for sales in Daimler-Benz's newly created commercial vehicles division. Schrempp would be appointed deputy director of this division on the board of management only nine months later.

When Liener became financial director, as part of a reshuffle at the top of the company, Schrempp was allocated a new boss: Helmut Werner. Werner and Jürgen Schrempp would from then on lead the commercial vehicles division—and matters developed quite differently from how the two of them had imagined.

Conditions were "catastrophic," says a high-ranking Mercedes employee of the management duo's performance. "There was that chaos at the Wörth works . . . eight out of ten trucks broke down when they were driven long distances." Even if such statements seem exaggerated, many things were not going according to Schrempp's plan. Even under the former commercial vehicles director, Gerhard Liener, annual sales collapsed from $10.7 billion to $9.5 billion, and there was already sufficient sand in the works in the truck and transporter sector.

At least they could record an expansion in sales to $10.4 billion and $12.2 billion respectively for the following two years. Sales rose from 234,000 to 258,000 commercial vehicles during the short Schrempp era, which could principally be put down to the "lively demand" on international markets. Schrempp was able to announce the "full utilization of capacity" at German commercial vehicle plants, with a growth in production of 6.7 percent despite the stagnation evident on the domestic market.

Neither Werner nor Schrempp could be credited with clearing up the trouble, though it would be unfair to blame them for the inadequate development of business on the German market in view of the short duration of their involvement. In this muddled situation Schrempp was "clever enough to simply give up"—that is the retrospective analysis of someone with intimate knowledge of the scene, who was himself active in Dasa's management. The prospects for the coming years were clearly ambiguous: The commercial vehicles sector would remain a headache for Mercedes-Benz AG for a long time to come. Another factor in Schrempp's choice to leave the field at the end of 1988—after only two years' activity in his area of expertise—was that maintaining the unsatisfactory status quo was not fulfilling enough for him.

The rationale given for naming Schrempp chairman of Dasa sounded logical enough: Schrempp had proved his staying power at Euclid and in South Africa, so it was an obvious decision by the chairman of the board, Edzard Reuter, to nominate Schrempp for the top post at Dasa; Schrempp's positive character attributes were the decisive factor. According to this official version, the decision in his favor came as a complete surprise to Schrempp.

The truth was that Edzard Reuter needed someone loyal and obedient to actively uphold his concept of an integrated technological concern, both within and outside the company. A Mercedes manager was more suitable for this than an MBB chairman whose horizons did not extend beyond aviation and aerospace—namely, Schäffler.

Schäffler's nickname, "Screwdriver Joe," implies an absolute aviation specialist. This nickname, however, also indicates that Schäffler's strengths were to be found more at a technical level than in managing the aviation giant that was to be created. In order to strengthen this impression within the company, it was also said of Schäffler that he was "a weak decision-maker" and a pure technician.

With all this working against him, MBB's boss blew his last chance by picking an argument with Edzard Reuter. Schäffler is considered a man of the older generation of aircraft, someone who thinks in classic Messerschmitt terms. For him, "an aircraft need not be economical, it just needs to fly," according to one source with direct contact. For Reuter, however, himself no aircraft specialist, the fact that "the thing has flown on a PC"—that is, computer simulations have indicated that everything will work—was sufficient. But when the chairman of the board confronted Schäffler with strategic questions, Schäffler replied in a lecturing tone—which made Reuter really mad. "Reuter then booted Schäffler out," recalls a close observer. Like Werner Breitschwerdt before him, Johann Schäffler was kicked upstairs to the squad of great technicians—to Schrempp's advantage.

In contrast to Schäffler's technical expertise, Schrempp's familiarity with defense technology was limited to the knowledge he obtained during his national service in 1968 and 1969, at a time when the U.S. Air Force was defoliating the forests of Vietnam with Agent Orange and bombing its villages with napalm, the Rolling Stones were singing "Street-Fighting Man," and youths with shoulder-length hair were experimenting with revolution. None of this was of much interest to Schrempp. He had his hair cut, did his basic training, and went on to become an armored infantryman. Later, as a technical instructor, he trained his comrades-in-arms on the hydraulics and electronics of military vehicles. Schrempp would have become a sergeant if he had stayed on.

His liking for parachuting, which he still indulges, stems from his time in the air force (Schrempp changed from infantry to air force sometime in 1968 or 1969). His inner drive impels the former top sportsman—who as a youth could run the hundred meters in eleven seconds—to continue pushing his physical limits. So he was delighted about the prospect of flying in a special two-seater training version of the Eurofighter 2000 at Mach 2.2 and with a force of nine Gs in the cockpit. Receiving his pilot's license after a successful physical checkup, the fifty-three-year-old was promised just such a flight by his British business partners on the project.

So all in all, wasn't Schrempp's knowledge of military and defense matters, in addition to his excellent physical constitution, enough to make him head of Deutsche Aerospace?

In any case, Reuter had made his decision: His protégé was to become chairman of Dasa, and Schäffler would have to accept being his subordinate.

Schrempp was named chairman-elect of the board of the newly founded Deutsche Aerospace AG on January 1, 1989, and Johann Schäffler was named as his deputy. At the next shareholders' general meeting, in May 1989, Schrempp took Schäffler's place on Daimler's board of management, marking the beginning of the end of Johann Schäffler's career—the same man who once had been one of the most powerful figures in the most powerful aviation company in Germany.

At his first interview with the press as the new head of Dasa, Schrempp was challenged with the question of whether he could imagine himself as Schäffler's successor. Realizing that he had been confronted with the most important question of all, Schrempp adroitly avoided mentioning any ambition toward the company crown. Instead he said that "this excellent man"—speaking of Reuter—"has bestowed the task of a generation" upon him. Schrempp indicated that he was ready to complete this assignment, and "what lies before me would take many, many years to fulfill."

Between the lines, however, he was saying that if he succeeded in accomplishing this "task of a generation" and developed Dasa into a flourishing concern, then he would be in a better position than his three main rivals on the board. Financial director Gerhard Liener was only four years younger than Reuter. And even the vice chairman of the board, sixty-four-year-old Werner Niefer, no longer appeared to be a serious opponent, as he was to leave his post at the shareholders' general meeting in summer 1993. The main competitor was Helmut Werner— and he was made of sterner stuff.

The detour via Dasa was more risky, but if it should succeed, then Schrempp would ultimately emerge from Werner's slipstream to reach the top far more quickly than by taking the classical route of the influential Mercedes faction.

Still the astonishment remained, however: Schäffler, the number two behind Schrempp? Even the experts were surprised by Reuter's decision. Schrempp, a commercial vehicle manager completely unknown on the aviation scene, taking over the management of this defense giant? What they did not know, however, was the extent of Schrempp's ambitions and how Dasa fit into them. Those who knew Schrempp from Pretoria were best placed to figure out that he would not be satisfied even by this board position—but they were in distant South Africa and not at Dasa's headquarters in Munich.

Schrempp had absolutely no idea of the ripples that would be created by the fusion of three fully autonomous companies, each exhibiting completely different corporate philosophies. If one ignores the sale of Euclid, Schrempp had never been in full control of the acquisition and amalgamation of companies. And there was no small difference between forging strongly different companies into a single effective concern and merging companies that have already worked in harmony with one another. Whatever the case, Schrempp was faced with a very difficult task. If he intended to survive, he would have to make some friends and dispose of some enemies—just as he had done in South Africa.

He began by tempting people to join his team, even though he was still director of commercial vehicles and had not yet formally shifted to Dasa. "Are you interested?" he inquired of Manfred Bischoff as the two sat in his office one day, but Bischoff was not sure how to answer. He took his time, looking around Schrempp's office. The relevant decisions were to be made there, he knew, and the choice Bischoff made that day would have a lasting effect on his life.

Bischoff had, only four months earlier, gone to Mercedes-Benz of Brazil, where he was managing director. But Schrempp's offer was doubly tempting: The economist could move up to financial director of Deutsche Aerospace AG instead of playing boss far away from home. The founding of the largest aviation and aerospace concern in the history of German business promised unlimited prospects for someone who wanted to achieve something with his life. And Manfred Bischoff had already resolved "not to have a normal career in the company" even before this all-important conversation with Schrempp in December 1988. He wanted to "build something new." And that was precisely what he was now being offered the opportunity to do.

Despite the tempting career prospects, the decisive factor that led Bischoff to agree was the impression the soon-to-be head of Dasa gave him: that Schrempp needed him. Bischoff gave it a little more thought and then told Schrempp his answer: "Then I'll do it."

Schrempp played a key role in the merger of Messerschmitt-Bölkow-Blohm with Deutsche Aerospace. After the German federal monopolies authority refused permission for the merger because of concerns about their possible domination of the market in the defense, aviation, and aerospace sectors, Daimler-Benz applied for ministerial intervention in reversing the monopoly commission's decision. Reuter sent his most able employee into the fray: Schrempp.

Nobody else could have gotten it done as quickly. In a twenty-five-page speech he made at a public hearing before personnel from the federal ministry of economics, the Dasa boss surpassed himself in economic analyses and sublime business policy promises. It was important "to develop the new technological areas vital for growth," to be active in branches "with strong growth potential," and to consolidate "development opportunities to the full" in the aviation and aerospace sectors.

And because it was not just a matter of the well-being of the company, but really also "our commercial responsibility," and because "the common good is also a yardstick of our activities," Schrempp concluded, the minister of economic affairs should be favorably disposed toward the plan.

Dasa chairman Schrempp threw all his political weight behind his appeals to the governmental decision makers to approve the merger contract, stressing "long-term job security" and "safeguarding the futures of thousands of workers."

He made the visionary claim that "the strategy of the Daimler-Benz Group was firmly focused on the next century." One could agree with this without reservation, for in this century Schrempp (with Reuter's complete support) took care of the job situation: More than thirty-five thousand jobs were "dismantled" between 1993 and 1995 alone, out of a total of eighty-six thousand employees—although, according to the official version, not a single employee had to be fired!

On September 8, 1989, the liberal minister of economic affairs, Helmut Haussmann, gave permission for the merger to take place. The "greatest administrative blunder in Germany's history" was complete, according to one of the many critical press editorials—not least thanks to the promises of Jürgen Schrempp.

The initial situation could hardly have been worse: a totally splintered German aviation industry consisting of a large number of competing and mostly ailing private companies; a public that was highly suspicious of the never-ending subsidies, amounting to billions of dollars, paid out of the federal budget; and the fierce competition within the Group to become the most powerful aviation manager in Germany. These extremely negative factors involved Schrempp in battles on many fronts. He responded by going on the offensive.

"We will turn MBB around in two years," predicted Manfred Bischoff, an optimistic type. Schrempp's response was typical of this most dynamic of men: "In two years? We'll get it done now." Bischoff acknowledges, without any envy whatsoever, that "Jürgen Schrempp has enormous drive."

Two years later, not much was left of the visionary concept of an integrated aviation concern. Dasa's reputation was unimaginably bad; the man at the top was battling against the vacuum left by the collapsing Communist threat; and, particularly after the dissolution of the Warsaw Pact organization, the relevance of military aircraft production was being increasingly loudly questioned, keeping Schrempp very busy.

Every new critical report increased the likelihood that Schrempp would be exposed as an incompetent "aviation expert" and that his deputy, Schäffler, would be given the chance to do better. Schrempp was forced to react before his competitor did indeed triumph.

This time he restructured the entire aviation and aerospace concern yet again. "The earlier concepts were simply wrong," suggests a manager at Dasa, adding: "We could not process all this."

It really was an achievement of a special kind when one looks back from the perspective of 1999, for in six years Schrempp came up with four different concepts for the structure of Deutsche Aerospace in the then called Daimler-Benz Aerospace, the DaimlerChrysler Aerospace of today.

That the years 1987 and 1988 were considered to be Schrempp's most economically successful on his way to the chairmanship borders on the grotesque. Even if one admits that political circumstances in South Africa were extremely difficult at that time and that the changes Schrempp introduced at Dasa did perhaps have a positive—if somewhat delayed—economic effect; even when all this is ascribed to Schrempp's account, the short-term balance is unsatisfactory: He was unable to claim personal success for the positive effects of the stringent adjustments he had introduced at Dasa before his departure.

Six Feet Under

There cannot actually be a better formula for success.
*Jürgen E. Schrempp, Chairman of Dasa's Board of Management,
on one of his new Dasa company structures in 1991*

Those on Schrempp's team had still not been able to bring the Dornier group, which had not bowed to Dasa's interests, into line. Schrempp's complaints about Dasa board member Helmut Ulke were as harsh as they were unwarranted.

Ulke was not a shy person. He was considered not only imposing and eloquent but also a farsighted and educated expert. A man like this had to be convinced first—which could not be achieved by a charming smile and the comment that he must, at long last, accept the new structural concept. He asked a lot of questions: "Isn't the Do 328 actually an excellent propeller-driven plane? And what about Dornier's medical technology? The market prognosis could not be better; are there not in fact exceptionally good future prospects?"

The Dornier boss's determined attitude grated on Schrempp. Ulke still dared to oppose him, and this was the real reason why Dasa's chairman took him to task.

One who experienced the conflict firsthand comments that Ulke "simply said a lot of good things—about flying's great prospects and the future importance of Dornier's medical technology." It really wasn't stupid, adds a board-level expert, referring to the Do 328's success today— at least in the jet version. At any rate, the company (today now part of U.S.-German aircraft manufacturer Fairchild Dornier) was making a profit, the number of workers employed was steadily rising, and it was extending its range of products. A stretch version of the Do 328, the 428JET, was being launched into the market, and its first customers were Lufthansa and Crossair.

It was all to no avail. Jürgen Schrempp did not tolerate criticism within his board for long. Ulke was certainly not Schrempp's most famous victim, but he held the dubious distinction of being the first one from Dasa's board. At the end of October 1991, Ulke's time on the board ran out. One day later a thoroughly loyal Schrempp ally joined the board in the shape of the new Dornier chairman, Werner Heinzmann. Ulke's demotion had been (as they put it so well in Daimler-speak) "harmonious."

Dornier, MBB, MTU, and Telefunken SystemTechnik, all brought together as Deutsche Aerospace, "have grown together to form an efficient union of companies," announced a self-satisfied Dasa board in 1991. The board's chairman, however, had to confess that the new organization of the management structure into four divisions—aviation, aerospace, defense technology, and engines—"could not yet be optimized as a result of the special legal situation in our companies."

Of course, he could point to a first success: In Dasa's second year of existence the company's balance sheet improved, thanks to Schrempp, from a loss of $74 million to a loss of $72 million, though sales dropped, too, from $6.75 billion to $6.65 billion. Thus 1990 was "above all characterized by the fundamental progress achieved within a framework of reduced tension between East and West," according to Schrempp, who nonetheless showed some disappointment: With the invasion of Kuwait by Iraq, "many expectations were again dashed." The expectations of German business regarding the hoped-for innovative impulse that was to be brought about by the union of aviation and aerospace under one Dasa roof were also dashed.

Had he been asked at the time, "What now, Mr. Schrempp? Have your balance sheets improved? Have your promises worked out?" Schrempp would have had to admit that they had not. But he was clever enough to point a timely finger at the company's defense and civil systems division, headed by Gerhard Jäger. Its sales had gone down again, and the number of orders received had fallen "unexpectedly sharply," according to the board's criticism. Was it not logical, then, for Jäger to have to leave the board of Deutsche Aerospace in July 1992, nine months after Ulke?

Events proceeded very rapidly again. Dornier's chairman, Werner Heinzmann, was named as Jäger's successor for the directorship of defense and civil systems in September. Heinzmann thus combined Ulke's aerospace division and Jäger's defense technology division in a personal union.

Only five months later, Johann Schäffler's last hours on the board had come and gone, too. Schrempp, the leader of the wolf pack, was ruthlessly clearing out his territory.

Of course, he could point to one further success: In Dasa's third year of existence the company's balance sheet really did improve, from a loss of $72 million to a profit of $26 million. Even if sales had dropped again, from $6.65 billion to $6.54 billion, Schrempp's efforts at last appeared to be bearing fruit. The euphoria, however, was short-lived, for only one year later the concern was in a worse mess than ever. "Dasa only made a profit in 1991," notes Manfred Bischoff; otherwise, it was still an extremely sickly child.

Hercules of Ottobrunn

The Board of Management was aware, right from the start,
of the Herculean task involved in forging Dasa
into a competitive business unit.
*Edzard Reuter, Chairman of the Board of Management,
to the Supervisory Board in November 1993*

One of Schrempp's most brilliant gambits was appointing a consummate professional, Detmar Grosse-Leege, as his personal PR manager. It was no coincidence that Schrempp was able to dispose of one rival after the other with few repercussions, as behind him was an intelligent figure who knew how to handle the media. The names Ulke, Mehdorn, and Schäffler also remind one of the excellent press control around Dasa's chairman.

As was proper for the Dasa duo of Schrempp and Schäffler, they occasionally did the rounds advertising their ailing company. The division of tasks during their joint excursions to the Ministry of Defense in Bonn or the government procurement office in Koblenz was clearly defined: "Schrempp got the formalities out of the way and Schäffler did the rest," reports one observer at Deutsche Aerospace. The only problem was that Schrempp was the boss of Dasa and Schäffler his deputy.

It was best not to tangle with Schrempp in such matters. And, according to former members of Dasa's management, Schäffler himself was doing everything possible to undermine his own position. Reports repeatedly surfaced in the media about how Johann Schäffler did not feel good about his situation and that Dasa's number two wanted to leave before his contract expired.

At its first meeting in Ottobrunn (in 1992), Dasa's supervisory board acceded to Screwdriver Joe's "request" to release him from his duties on the board at the end of the year. The New Year's Eve celebrations at Schäffler's home were correspondingly subdued this time. Schäffler had already lost his job as head of MBB in September, and the vice chairman would be formally pensioned off, with expressions of thanks, on December 31, 1992.

Schäffler out, Mehdorn in. Schrempp plugged the hole left by the specialist Schäffler with the aviation expert Hartmut Mehdorn (he had been chief of the largest defense company in Germany—MBB—before founding Dasa). This was a clever gambit: Mehdorn was considered to be an aviation professional and supplied the expertise that had so distinguished Schrempp's previous deputy. Like Schäffler, he was the top technician on the board's team, who—unlike Schrempp and his financial director, Bischoff—knew how an aircraft works. Better still, while as enthusiastic as Screwdriver Joe, Mehdorn also knew how to milk Bonn's financial cows. He made heavy demands of the Ministry of Defense, and "very much improved harmonization, coordination and targeting of research and development policies between the industry and the ministries responsible." Mehdorn fought for "greater political support of international marketing activity" and wanted a "one hundred percent output quota" for Dasa's aviation division and "a budget provision for technological projects" totaling up to $180 million a year. Hartmut Mehdorn's opinion counted in Bonn, something that impressed Jürgen E. Schrempp tremendously.

Dasa was a long-running drama at both of Daimler's boards. Hardly a meeting passed without Edzard Reuter or Schrempp having to inform the management committees about a further deterioration or another unsuccessful endeavor. The increasingly dramatic developments should have caused the board of management to raise an outcry and the supervisory board to agitate against the Dasa disaster long before. That this was not the case was solely due to suppression by Daimler's chairman, so blinded by his cherished model of an integrated technological concern, and by the chairman of the supervisory board, who simply looked on, largely inactive to the bitter end.

Schrempp could do what he liked: None of his structural concepts proved effective enough to successfully withstand the worsening circumstances. The record sales figures of 1993 (entirely due to the Fokker deal) were followed by a steady decline from $9.6 billion to $7.55 billion in the next two years. During his last three Dasa years alone, Schrempp frittered away $2.8 billion. Then, in the summer of 1995, Schrempp made his move toward Möhringen.

At the supervisory board meeting on November 3, 1993, members of the management and supervisory boards observed a minute's silence at the instigation of Hilmar Kopper. The vice chairman of the board, Werner Niefer, had died three weeks earlier. And it seemed that Niefer's death would leave its mark on the hours to come. Walter Riester was attending his first Daimler supervisory board meeting, and it would be a long time before the metalworker would accustom himself to the climate prevailing at this committee.

"As you know, Dasa was founded only a few years ago by putting together numerous larger and smaller companies." Edzard Reuter allowed himself time to present the reasons for the very upsetting situation in copious detail. "These companies were, however, not very compatible." This was so, for example, in the case of "MBB on the one hand and Dornier on the other," Reuter said, formally covering for Schrempp while having to justify the concept that he and Schrempp had worked out together.

The board was aware, right from the start, of the "Herculean task" involved in "forging together an internationally competitive business unit" out of Dasa, continued the head of Daimler. There was no question about it—Reuter was absolutely correct. And one of the supervisory board members should have spoken out, for all those who had been members of this committee for the last four years had had to sit through this litany of concealment and deflection at almost every meeting. Reuter, as the Zeus-like figure, was allowed to continue without interference and Schrempp, as Hercules, could make full use of his methods of retaining power without hindrance.

The fate of Helmut Ulke really should have served as a warning for Mehdorn. Those who know the man who was at the time Dasa's chairman know that Schrempp loves being contradicted—as long as he winds up being correct in the end. Basically Schrempp believes that he always has the better arguments. Mehdorn, on the other hand, was a freethinker, sometimes critical and always precise in his analyses. He viewed the acquisition of the stake in Fokker with great skepticism right from the start. Nevertheless, he felt that he had to go along with it because, ultimately, he could not be director of Dasa's aviation division and simultaneously publicly criticize his boss's decision to merge with Fokker.

The split with Mehdorn became inevitable when the Fokker case developed into Schrempp's worst-ever defeat. Schrempp had to admit to his billion-dollar mistake openly, and a confident head of aviation had always prophesied that this would happen.

As if this were not enough, Hartmut Mehdorn also contributed toward making the personal relationship between the two of them reach ice-age temperatures: In his view Schrempp was totally ignorant of the finer points of avionics. It was completely predictable that Schrempp's level of empathy with Mehdorn continued to tend toward zero. The rest could be taken care of.

Reuter got carried away in his enthusiasm. Everyone involved had taken on the challenge of "tightening up the division of responsibility" and "creating a healthy economic basis" for the company's sites. A neutral observer, knowing nothing of the catastrophic balance sheets of this thoroughly lackluster aviation concern, could be forgiven for mistaking this for the end-of-year celebrations of a thriving family company. Reuter sent out messages throughout the company encouraging perseverance.

As superficial as the statements continued to be, he skillfully spread the responsibility to include everyone. "You have always stood by our basic principles," Reuter said in his situation report in November 1993, expressing his gratitude to the supervisory board. And here, too, he was absolutely correct. Even the employees' representatives had always voted with the shareholders, with the one exception being the merger with MBB.

Over the breakfast table on Monday morning, Hartmut Mehdorn opened the Bavarian political magazine *Focus* and read what an incompetent Dasa aviation division head he really was: "Schrempp is still waiting for forward-looking strategies from his section head for regional jets and turboprop aircraft. . . . Schrempp fears that unless a new sales concept is created quickly," Dornier's regional aircraft, the Do328, which cost $500 million to develop, would never turn a profit. And the magazine claimed that as Dasa's top man was in no mood for taking the blame for the continuing run of failures, "the company would make Mehdorn responsible for its lack of success." "Thus Hartmut Mehdorn is now under pressure at Deutsche Aerospace," commented the influential magazine about the top floor's attitude at Dasa headquarters in Ottobrunn. Mehdorn must have completely lost his appetite for the breakfast in front of him as *Focus* informed him that he did not know how to delegate and "buries himself in detail." Actually, he was "far too seldom to be found at his desk, complains the chain-smoking Schrempp."

That line was a bit rich, for Schrempp himself used every opportunity to leave his unbeloved desk to visit a factory or take a trip abroad.

In a friendly move, the *Focus* article also tipped off the still-presiding aviation director that "he has until September" and that his successor was already waiting in the wings.

This sort of foreshadowing was common practice for Schrempp, and when such media reports appeared, it was widely known who was pulling the strings in the background. Usually, but (as in this case) not always, the bad news appeared in *Der Spiegel*. This time, however, it was not possible; after all, Hartmut Mehdorn himself had good connections at *Der Spiegel*'s Hamburg headquarters. Thus in this case it was *Focus* that provided inside information to an interested public, and Mehdorn was presented in an increasingly poor light during the coming months. He lost one post after another and was eventually relieved of his seat on Dasa's board on September 30, 1995.

Hartmut Mehdorn's fall did not work out quite as smoothly as Schrempp would have liked. Sales of the Do328 took off in 1994, as Schrempp was forced to admit in May of the following year. The short-haul aircraft had gained 40 percent of the U.S. market; both firm orders and options were buoyant. Despite this brilliant success, in July 1995 Mehdorn had to hand over his seat to Dietrich Russell. Schrempp and Mehdorn, whose contract would ordinarily have continued until 1997, could not even come to any agreement on the otherwise customary formulation announcing his "amicable" departure.

Once again—and this is something that Schrempp's colleagues had already learned—Daimler's chairman did not easily forgive.

Schrempp was elected the new chairman of Daimler's board of management at the board's meeting in June 1994, in gratitude for his convincing performance, and at that time he also swapped his desk job as Dasa chairman for the post of chairman of Dasa's supervisory board (though he continued to keep an eye on the aviation and aerospace giant). Shortly before this meeting, Hilmar Kopper indicated to Edzard Reuter that his move to the top of the supervisory board, contrary to the original promise, would now meet with "considerable resistance from the shareholder representatives." Ultimately, Reuter's nomination was thwarted by Kopper. But this decision really did not have anything to do with Schrempp.

When the first opportunity to choose a German to head Airbus production in Toulouse came up in August 1997, the French side signaled that they were prepared to compromise. Hartmut Mehdorn's chances were considered good after the French government awarded this "great European" the Aviation Medal. Jean Pierson, former chairman of the Airbus consortium, could be elected to the supervisory board and Mehdorn could become chairman of the board of management. This, however, would require the enthusiastic backing of a Daimler board member and the chairman of Dasa's supervisory board—who would have to be willing to consign their old argument to history and recall former Dasa board member Mehdorn back from Heidelberg.

There were various reasons why Hartmut Mehdorn would ultimately fail to become head of Airbus. One of the most weighty was that Schrempp preferred Manfred Bischoff as chairman of the Airbus supervisory board to Hartmut Mehdorn as chairman of its board of management. That Bischoff was a worthy successor to Reuter was certain, but his selection was no coincidence, for Schrempp, true to his character, would do everything in his power to promote his friends—and hardly ever abandoned a grudge against his adversaries.

At Dasa too, as in South Africa, Schrempp demonstrated that he was not one for making smart personnel decisions. Those who were not a hundred percent for him, and even those who dared to argue with him, could expect continuous flak from Schrempp—and nobody survived that. Either the supposed opponent got himself into line pretty damned quickly or else he left and could then—mostly provided with respectable settlements and in "mutual harmony"—look around for a new field of activity. Ulke, Schäffler, and Mehdorn had their names put on the long list of those who fell by the wayside during Schrempp's rise to the pinnacle of Daimler-Benz AG.

Apart from 1991's meager $26.6 million profit, the balance sheets had been uniformly in the red. In the end the financial balance of Schrempp's rule from 1989 to 1995 was a total deficit of $3.1 billion.

Why, however, was no serious pressure exerted on Dasa boss Schrempp? Why did no heavyweight Daimler representative demand his resignation? There were two principal reasons for this: First, after the (not entirely voluntary) departure of the only serious contender, Schäffler, there was no reasonable alternative. In addition, the Group's chairman had swept all the failures under the carpet of his vision. If Edzard Reuter had himself introduced Schrempp's demand for a 12 percent minimum yield from each of the Group's divisions before Schrempp had secured his seat as chairman of the board, then Schrempp would hardly still have been a Daimler-Benz AG employee. But for years Reuter personally covered up for the man he had appointed to be Dasa boss. If he fired him, he feared, then his dream of an integrated technological concern would be disintegrated. Reuter was so dazzled by his personal fantasy that he underestimated Schrempp. Edzard Reuter meanwhile may have come to regret this lost opportunity.

Looking back on the six years of his leadership from May 1989 to May 1995, Schrempp reaped few kind words within the company, principally as a result of his lousy performance. "Look at Bischoff. He has style and talent—and he's knocking Dasa into shape" was simultaneously praise for Bischoff and criticism of his predecessor, Schrempp. This kind of statement, influenced by reservations about Schrempp, was heard again and again when one talked to Dasa employees. As popular as Schrempp appeared superficially, the commentaries heard off the record were critical of him. Even today many still consider Schrempp to be a trucker with little or no idea about aircraft construction.

To be fair to Schrempp, his Dasa years were affected by extremely unfavorable factors: the collapse of the East-West conflict, the absence of an enemy after the dissolution of the Warsaw Pact, the resulting reduction in the amount of investment from the defense budget, and the long-lasting weakness of the dollar at that time. All of these elements contributed to a situation in which Jürgen Schrempp could survive only through extensive lobbying and with government assistance amounting to billions.

But this was only part of the truth. Schrempp's repertoire of explanations is based on the notion that he was caught off guard by all these nasty surprises. That critics, and not just from outside the Group, gave early and insistent warnings against the diversification strategy of Edzard Reuter and his protégé Schrempp was, however, something that neither Schrempp nor Reuter was prepared to accept.

At no time was Schrempp able to turn Deutsche Aerospace into a consistently profitable company. The record deficit of 1992 doubled to $369 million the following year. The 1995 fiscal year closed the six-year era of Dasa chairman Schrempp with another record loss of just over $1 billion. Schrempp did not shy away from injecting another huge amount of money into the concern, increasing the deficit for that one year alone to nearly $2.2 billion. This includes the extraordinary expense of the Fokker acquisition. Is this all to be ascribed to Reuter?

Let's take another look back. After his loss-making reign as MBSA boss, his failure to restructure Euclid Inc., and the heavy losses resulting from his six years as chairman of Dasa, the red carpet was rolled out for him in Möhringen: Schrempp was crowned Lord of the Stars in the summer of 1995. There were many explanations for this—but on no account could they include an even mildly successful economic performance during his reign.

Or has Schrempp always moved on too soon, before his labors have borne fruit? For the situation looks very different now that various important markets have recovered. After Schrempp's departure, Dasa made an operating profit of $296 million that would increase to a noteworthy $649 million in the following year.

The European aviation concern is a highly successful business today. After 460 firm orders for Airbus in 1997, its head, Noel Forgeard, was able to announce orders for as many as 556 aircraft in1998. Headlines in the business press proclaimed "Airbus Soars to Record Results." Is this all to be ascribed to Schrempp?

Archrival Boeing in Seattle also managed to increase sales in 1998, though far more modestly. This industrial leader suffered enormously from the effects of the Asian crisis, and announced a further 7,000 job losses in the summer of 1999. Thus 12,000 of its former workforce of 231,000 have already been fired. And the situation is hardly likely to improve before 2001, according to statements made by Boeing.

In contrast, champagne corks were popping to welcome in the new millennium at Airbus. "The increase in Airbus production in Germany was sensational," announced the jubilant and self-satisfied Dasa spokesman Rainer Ohler, adding that in 1999 they had a backlog of orders for more than 1,300 aircraft still to be delivered—at a value of more than $92 billion. So can all *this* be ascribed to Schrempp?

There would have to be mergers if Europe's aviation, aerospace, and armaments industries wanted to survive the all-powerful American competition. Schrempp set his sights correspondingly high: He wanted to set up a powerful "European Aerospace and Defense Company" (EADC) together with the other European defense firms to oppose the superior strength of the U.S. companies Boeing and Lockheed Martin. This would have been, at least by European standards, an enormous business, with annual sales totaling $32 billion and no fewer than 130,000 employees. Naturally, according to Schrempp, Germans would have a place in the cockpit.

Shortly before Christmas Eve 1998, the merger of the British aviation giant British Aerospace (BAe) with DaimlerChrysler Aerospace was considered to be only a matter of days away. Then, however, resistance surprisingly surfaced within Britain's major industrial sector. During the Christmas break Lord Simpson, managing director of the mixed General Electric Company (GEC), made BAe's chairman, Sir Richard Evans, a lucrative offer that he could not refuse: Simpson offered to take over the attractive GEC defense electronics division under the name of Marconi, and Evans agreed.

This decision hit Schrempp like a bolt out of the blue. With this announcement, the possibility of the amalgamation of more-or-less equal partners that he had called for had collapsed. After all, the German top managers had agreed to settle for a share distribution of 60–40 in British Aerospace's favor.

Dasa boss Manfred Bischoff saw the same consequences: A European solution had to be based on an "equal balance of industrial codetermination . . . otherwise it wouldn't work." With the Marconi takeover, Daimler-Chrysler's share would sink to about 20 percent, and Schrempp's influence would be marginalized.

The collapse of the deal between Dasa and BAe was ultimately "simply because of the management question," according to a deeply frustrated Manfred Bischoff. Bischoff had assumed that if there had been two top positions, one would have been occupied by the Germans. After all, Dasa was not about to "sell out German interests" and would "therefore not surrender to the dominance of others."

And Schrempp certainly would not. He had personally negotiated with Sir Dick Evans himself in what he described as an excellent atmosphere. Thus with great disappointment the DaimlerChrysler chairman announced at a spring 1999 press conference that the European aviation and aerospace giant "would not be created" as an EADC "in one large success." And Jürgen Schrempp described the realities more clearly than any other European armaments manager: "I think that this dream is over."

The result of all this will probably be that the harsh competition between the defense companies will be fought out more fiercely than ever. "There is only room for one aviation and defense company in Europe," BAe's chairman Sir Richard Evans said, issuing an undisguised challenge.

For the Germans, it was clear that the coming years would involve fighting on two fronts: retaining industrial superiority in European aircraft construction, and battling for global air supremacy against seemingly overpowering competition from the United States. Schrempp accepted the declaration of war and struck back just a short time later.

In the second week of June 1999 "the news was all over town," according to a member of Daimler's controlling committee, that negotiations with the Spanish aviation company Construcciones Aeronauticas SA, known as Casa, were in their final phase. It was undoubtedly an attractive takeover candidate, with a record profit in 1998 of $49 million (a 20 percent increase over the previous year), and all four of Europe's large aircraft manufacturers (British Aerospace, Aerospatiale Matra, Alenia, and Dasa) had expressed an interest in Casa at the beginning of 1999.

But members of Dasa's supervisory board—specifically those representing the employees' side—were furious that they had not been informed ahead of time and had to "find out what's going on from the news on television." This was despite Manfred Bischoff's promise that any decision on Dasa's future, including a merger with another aviation company, should not and would not be made without the involvement of the supervisory board. In exchange, the employee representatives promised that they would take part in any meeting, however short the notice given; this was intended to avoid giving Dasa's chairman of the board or his personnel director, Hartwig Knitter, an excuse for not including them. Yet the Dasa-Casa action showed what Bischoff thought of the supervisory board members.

That Manfred Bischoff informed the controlling committee—which was supposed to be in control of his actions—only when he deemed it necessary, and not when they wanted to know, was not particularly surprising. On the other hand, the fact that, even one year after the Daimler merger with Chrysler, not all the members of the supervisory board (even within the employee representatives' camp) knew each other by name was extremely disturbing. As long as the unionists' cooperation failed to extend across the Atlantic, Manfred Bischoff and Jürgen Schrempp would have an exceptionally easy time.

This was not how things were supposed to work, of course. Upon receiving a seat on the supervisory board after the merger, United Auto Workers (UAW) president Stephen P. Yokich had declared, "I am really looking forward to working on the supervisory board," and his enthusiasm was evident. He added that "the participation of German workers is a great thing," for it allowed continuous dialogue between employees and management and, he said, meant that all sides of the company could react in unison to new developments.

Nevertheless, on June 11, 1999, Schrempp, as DaimlerChrysler CEO, announced the signing of a declaration of intent for the merger. He had every reason to celebrate. The Dasa-Casa merger was a joint decision, not a one-sided takeover—an especially pleasant situation following their negative experience negotiating for Marconi. The Dasa-Casa amalgamation created an aviation and aerospace concern with fifty-three thousand employees and annual sales of $10.2 billion. Even more important was Casa's participation in the joint European project to build the Eurofighter (formerly called the Typhoon) fighter plane and in the European Airbus consortium. Dasa-Casa now become the largest participant in Airbus, with 42.1 percent.

Furthermore, the Spanish had gained a reputation for being "easy partners." "We would retain the know-how," according to a leading Dasa representative, who privately considers the Spanish as simply an "extended workbench" for Germany's production line.

The litmus test would be where the world's largest aircraft, the Superjumbo A3XX, was built: in Spain or in Germany? And although the Spanish government had guaranteed to build the roads providing the direct access to the sea necessary for shipment, Spain's chances deteriorated with the Dasa-Casa merger—unless the Spanish made it worth Schrempp's while. This would, however, have required further subsidies in the millions. An insider (who, hardly surprisingly, does not want his name mentioned) objects, "Some of those at the top have become very rich by doing business in shares."

All in all, the Germans are quite happy. Says a delighted Bischoff, "Together we can take the largest international step ever toward reorganizing our industry in Europe." He continues, "The consolidation and fine-tuning" of "large successful programs such as the Airbus or the Eurofighter" can now be realized.

No wonder, then, that Jürgen Schrempp speaks of a "great success" and says he detects an improvement in DaimlerChrysler's global position through the joining of the two concerns that would open up new opportunities for the future. Coming from Schrempp, what such statements portend for the American competition is clear.

Whether Jürgen Schrempp succeeds in putting together the largest, most powerful European aviation and aerospace concern or whether the

British and American competition ultimately come out on top will become apparent in a few years' time. If DaimlerChrysler's CEO fails, then his successor will have to soberly analyze the books.

However, at the present time, others besides Schrempp are jubilant.

A self-satisfied Wolfgang Piller, a Dasa board member under Schrempp and today president of the Federal Association of the Aviation and Aerospace Industries (BDLI—a lobby representing the interests of the aerospace industry), points out that three-quarters of all civil and military aviation and aerospace production in Germany was carried out in the works of DaimlerChrysler Aerospace. This includes a wide range of assorted goodies for democrats and dictators all over the world, from antitank mines to combat helicopters and the new Eurofighter. And Schrempp sees no very serious obstacles to arms exports, even to those governments that do not honor human rights, such as in Turkey or Indonesia—though in the end he has done only what German export regulations permit him to do.

The celebration about Dasa aside, there is some criticism of Daimler-Chrysler in terms of the weaknesses in automobile production on the Daimler side. Thomas T. Stallkamp, who holds the number three position and is not known as a man for superficial analyses, points out that while Chrysler is a "very large, very homogeneous company that builds 3.3 million cars a year," Daimler produces "far fewer" and, in addition, was "not as profitable." What such statements ultimately mean for Dasa's survival is obvious.

The Lord of the Falling Stars

I am not sure that the man at the top is really so important.
Morris Shenker, Chairman of Mercedes-Benz of South Africa

He happened to sit next to her at the German headquarters' visitors' cafeteria, and he fell into conversation with her more or less by chance.

"I'll write my name out for you," Jürgen Schrempp said, and did so, complete with the initials J.E.S., on the placemat in front of him. Martine Dornier-Tiefenthaler noticed he placed particular importance on the middle initial. The *E* stands for Erich, the name of his father's eldest brother, Jürgen's godfather.

She found this new acquaintance "great, because he's uncomplicated." Her first impression was positive, and above all she was impressed by his directness and openness.

There was no doubt that Jürgen Schrempp was quite different from the mass of people.

"He was really enthusiastic. He was wide awake and took in what was going on around him," Dornier-Tiefenthaler says. As serious as the words are intended to be, they sound ironic given the fact that this lawyer is considered by many to be Schrempp's number-one enemy.

Schrempp's defenses go up when her name crops up. "She had a great success once—1988," says Schrempp dryly, referring to the contractual negotiations between his predecessor and Dornier-Tiefenthaler, acting on behalf of the joint heirs of the Dornier aviation and aerospace company with sparkling success. When Daimler's chairman wanted to fully incorporate the company, based in Friedrichshafen, Germany, into Dasa at the end of the eighties, Dornier-Tiefenthaler, a crafty legal expert, threatened to break off talks—and ended up getting what she wanted, to the consternation of Edzard Reuter. The contract she negotiated gave nearly $31 million to the family, as partners in the business, and a guaranteed minimum 15 percent dividend on Dornier's shares—for life, and independent of the financial performance of the company.

Martine Dornier-Tiefenthaler matter-of-factly acknowledges that she may be the "token woman" on the Mercedes-Benz supervisory board, but she also sees herself as a potential "new beginning." From Schrempp's point of view, however, it was a beginning that started out badly. When in a February 10, 1996, interview, she criticized the sale of Dornier's medical technology section—"a market of the future," in her opinion—and strongly attacked Jürgen Schrempp, who was responsible for the sale, Daimler's boss blew his top.

In a furious ten-line letter, which Dornier-Tiefenthaler read about in a newspaper while she was on a ski trip in the Alps, Schrempp informed her that he could see "no more possibility" of "any further sincere cooperation" on the Mercedes supervisory board, and for this reason he had arranged for "her dismissal from the supervisory board" to be proposed at the forthcoming shareholders' general meeting.

The shadow boxing developed into an open exchange of blows. And in the process Jürgen Schrempp was forced to face the unpleasant fact that even the Lord of the Stars' power was limited. Two days after Schrempp's short and succinct letter, he read a lengthy reply in the newspaper *Stuttgarter Zeitung*.

With her dismissal from the Mercedes-Benz supervisory board, Dornier-Tiefenthaler charged, Schrempp had deviated "once again from the fundamentals of the decision-making process of Daimler-Benz AG." And what interested her "as a lawyer" was the question of "a board of management member's behavior if he wants to dismiss a member of the supervisory board over differences of opinion about the Group's business policies." Ultimately, "good conduct toward members of the board of management was not a duty that the law [demands] of a member of the supervisory board." And in order to really annoy the man in charge of the Group, Martine Dornier-Tiefenthaler—contrary to the usual practice in such cases—postponed her decision about whether to resign from the post, for she simply did not want to spare Jürgen Schrempp the "embarrassment of having to dismiss me."

Schrempp insists that it was only through the constant attacks on Reuter and himself that she gained media attention.

Nevertheless, this woman got to him badly. His battles with this experienced lawyer, carried out over a number of years in a variety of venues, demonstrated that the brightness of the star above the Daimler-Benz headquarters could lose much of its luster—and that the Lord of the Stars could quickly become a fallen star.

When asked if Schrempp is macho, Martine Dornier-Tiefenthaler laughs and replies, "Of course, and how!" She also counters with her own question: "What normal person goes up into the high mountains without any breathing apparatus? The man has to prove himself physically." She cannot help but be ironic when she philosophizes about Schrempp's disposition. "His whole manner is really just an imitation of how he thinks some little yob would behave." She does not mean this in a nasty way, but she grins smugly.

Whether friend or foe, admirer or critic, all agree that when Edzard Reuter had to vacate his seat, with a little bit of help from his successor, Jürgen Schrempp was ready to step in. He was the right man at the right time, in Dornier-Tiefenthaler's view, but that does not make him a god. With this opinion she sets herself apart from the army of those who idolize the chairman of Daimler's board of management—or at least act as if they do.

A job such as that of the chairman of Daimler-Benz AG, says someone who knows Schrempp well, is "much bigger than a single person" could fill, "just as was the case with kings and queens." It was simply that "the public believes that there really are some people who can manage the job."

Such views accord with those of Morris Shenker, who, when after a quarter century at the top of Mercedes in South Africa was asked how the company would manage without him, just waved the question aside. He said he found the compliment very nice, but "when you have to produce and sell a product like Mercedes-Benz," he observed, the man at the top is not so important.

It was no accident that Deutsche Bank—DaimlerChrysler's largest stakeholder—transferred Schrempp from the post of chairman of Dasa to that of chairman of the board of management of their showpiece company. What was needed at the time was a tough, radical restructurer, and in the mid-nineties Schrempp was indeed the right man at the right time. Hilmar Kopper's post was reconfirmed at the shareholders' general meeting in the summer of 1998. The Kopper-Schrempp duo were on the same wavelength, in a similar way to the earlier Herrhausen-Reuter team. And as long as the banker continues to watch over and protect Jürgen Schrempp, Schrempp will continue to have a free hand.

PART 3

The Globalizer

CHAPTER 7

Straight Capitalism?

We could not charge a higher price for our cars in the showrooms
just because we put a sticker on them saying,
Constructed with full sick-leave pay.
Jürgen Schrempp

When Jürgen Schrempp took over control of Daimler-Benz AG, he made a lot of changes in the way things were done. Unlike his predecessors, he discussed in depth key issues during the biweekly meetings of the board of management—late into the night if need be.

On September 24, 1996, the ten-person management team held their meeting, unusually, on the fringes of the International Commercial Vehicle Exhibition in Hanover. The members took their places around the table, and Schrempp has never forgotten this situation: "Mr. Werner sat on my left, Mr. Gentz on my right. Mr. Werner wanted the board to go along with the decision of the others." Schrempp was, in a sense, trapped between the head of Mercedes and the financial director during a meeting in which the main subject of discussion was not even on the agenda.

The chairman of the Corporate Works Council and deputy chairman of the Supervisory Board, Karl Feuerstein, had written in a letter four days earlier that the steel industry's employer association, Gesamtmetall, was encouraging its member companies to flagrantly breach wage agreements, thus carrying the dispute "to the factories." He warned that if this continued, the "relationship [that has up to now been] characterized by dependability and mutual predictability on both sides" would be "seriously damaged."

Feuerstein's threat was not to be taken lightly, and there would be serious consequences if a nationwide strike action was instituted.

"Where do the other car firms stand on this question?" asked Schrempp. The issue of a 20 percent reduction in wages in cases of illness immediately became the central topic. (In Germany employees get 100 percent of their wages if they are sick; Schrempp and other industrial leaders tried to install a new system where employees would only receive 80 percent. This attempt failed due to union protests.) Daimler's chairman was cautious. Implementation of the government model (which would allow, but not force, employers to reduce sick-day wages) would mean that the company's board of management would be the first to act in the interests of the employers' associations. And, after all, about 220,000 employees at Daimler-Benz alone would be affected by any wage reduction. As the federal government had just introduced the necessary legal framework, the new arrangement could be put into effect as early as October 1, just a week away. But Schrempp feared strikes that could last weeks and wanted to safeguard himself against this possibility.

Schrempp learned that Gesamtmetall had already met with the boards of Bosch, Siemens, and Porsche and that a decision in line with the recommendations of the government and the employers was expected. And Dieter Hundt, chairman of the Metal Industry Association of Baden-Württemberg, had announced that the new sick-pay policy would be implemented immediately in this state in the southwest of Germany.

Schrempp, still undecided, faced a dilemma. On one hand, he had been deliberately cultivating a close relationship with the union side and above all Karl Feuerstein, whose team on the supervisory board had unanimously supported his election as chairman of the board of management. On the other hand, Helmut Werner and Manfred Gentz were exerting a lot of pressure. The head of Mercedes was making insistent demands for the whole board to go along with the other automotive companies.

"All the committees had agreed before the board meeting in Hanover," Daimler's chairman explained. Approval to go along with the recommendation made by the metal industry's employers' association had indeed been obtained from all the Mercedes and Dasa board members. The decision was more or less made easy for Schrempp, so it was not surprising that after a short debate the board came to a unanimous conclusion to adopt the new policy, falling into line with the rest of the industry.

Schrempp did not foresee that the board decision would give the metalworkers' union an unprecedented boost—and that it would spawn the fiercest labor-management conflict in the hundred-year history of Daimler-Benz AG. But within a day Feuerstein had angrily reproached the company for taking a "leading role" in the matter, and IG Metall members initiated action just one day after the decision in Hanover. Spontaneous work stoppages took place at several Mercedes facilities, and very soon a wave of strikes had spilled over to other factories.

Feuerstein openly threatened that his union would take the toughest possible countermeasures following the "outright breach of wage agreements," suggesting that management imagine how things would be when "it is time for the extra shift and nobody turns up." What was just a threat on Wednesday soon become reality when IG Metall canceled the extra weekend shifts at Mercedes factories.

The massed anger of the affected workers focused on Schrempp. With his description of Schrempp as "the Rambo of the nation," Walter Riester, at the time vice chairman of IG Metall (now Germany's minister of employment) provided Daimler's chairman with an epithet that would stay with him for years to come.

At Mercedes' Untertürkheim works alone, ten thousand people demonstrated on October 1 against "the company Rambo," as they loudly proclaimed him. Sixteen thousand more put down their tools at the Sindelfingen factory. Sixty-three thousand Daimler employees—almost a third of the Group's entire workforce in Germany—were on strike only one week after the board's decision.

Furthermore, Heinz Kluncker, then the union boss, protested that Schrempp's demand for a reduction of wages in cases of illness was "callous capitalism" aimed against the autonomy of collective bargaining.

Schrempp was now without doubt the most hated manager in Germany. Manfred Bischoff reduces the situation to its bare essentials: "And suddenly you're the capitalist pig."

At nine o'clock in the morning on Tuesday, October 1, a plainly furious Karl Feuerstein stood before his army of "Benzers" and declared war on the enemy in Stuttgart. From his point of view the worst thing was the fact that "the largest industrial concern is leading the way." The mood among the thousands of metalworkers at the Mannheim Benz factory was explosive. "And Daimler-Benz was the first to infringe the law and breach wage agreements," raged the chairman of the central works council to a shrill hail of whistles. And he went on, "This has not happened since 1945 and is a declaration of war against all Daimler-Benz employees." Nonetheless, he was still one of the more moderate protesters. Only grassroots pressure pushed Karl Feuerstein into becoming a vociferous opponent of the Daimler board.

The incensed Mercedes workers were quite another case. Embittered, they give free rein to their anger on banners and by chanting in unison at the Waldhof-Mannheim works. "Those who break the law, reap the storm" read one placard. Feuerstein concluded his speech to the accompaniment of thunderous applause: "Colleagues, I am not permitted to call upon you to lay down your tools." But everybody knew what they had to do.

The unions had originally gained the right to full sick-leave pay four decades earlier (in 1957), after a strike that lasted 117 days, the longest in IG Metall's history. Further escalation was therefore a foregone conclusion. IG Metall planned a large demonstration in Stuttgart on October 10, and the culmination of their protest was set for October 24—forty years to the day since the start of the union's successful sixteen-week strike for full sick-leave pay. IG Metall was absolutely determined to stop Jürgen Schrempp. And the man on the eleventh floor of the Group's headquarters had still not recognized the danger he was facing; indeed, Manfred Gentz had issued a warning against illegal strikes organized by the union and had provocatively announced that "we will not be blackmailed."

While resistance in the factories was bringing about increasingly widespread repercussions, frantic discussions were going on behind the scenes. Schrempp was told that his image would suffer if he continued on his confrontational course; Manfred Göbels (chairman of the Senior Managers Committee, who usually sides with Schrempp) tenaciously worked on the chairman of the board, who still refused to give his approval for a meeting with the chairman of the central works council. Despite this, Schrempp was astute enough to make a few necessary phone calls. It was a quarter to six that evening when he called Göbels a second time. In the meantime, he had spoken to key figures in both Gesamtmetall and IG Metall as well as to his financial director, and all agreed to a meeting after Schrempp signaled that Daimler wanted to negotiate—knowing full well that he had lost this campaign, which had never been his to begin with.

Manfred Bischoff openly analyzes the mistakes that were made at that time: "It was a misjudgment that I, too, was guilty of." Even though Daimler's board members had wanted "to handle the topic very carefully," they were still "consistently in favor of wage reductions." The man who is now Dasa's chairman recognizes that the basic problem stemmed from the fact that the German government had created the framework for the change, and if the industry had not made use of it, "then we would have a problem." And that was precisely what Daimler's board had created for itself. All other companies retreated in fear of strikes, leaving Daimler and Schrempp standing alone. IG Metall organized the strikes against "Rambo" Schrempp, and Schrempp lost. Bischoff admits his own share of the responsibility for the development: "I blame myself for not having correctly appreciated the great sensitivity toward this topic on the part of the workforce."

Bischoff believes that the real blame did not lie with Schrempp. "The public's criticism of Jürgen Schrempp was unfair," he says, and he adds that he was "particularly sorry" for Schrempp, as Schrempp was "otherwise always at the forefront of the fight for basic rights in matters of employment." Bischoff goes on to say, with much emotion, "I suffered with him."

Was Schrempp simply too trusting? In his naïveté, did he blindly rely on the statements made by his colleagues on the board? And after such gross miscalculations, should there not also be personal consequences?

Whatever the answers, with his lack of resistance at the board meeting on September 24 he put himself firmly in the camp of those who wanted reduced sick-leave payments. It was interesting to note that his one-time rival, Helmut Werner, was among those who induced him to do it.

Manfred Bischoff was scathing about the U-turn by their comrades-in-arms on the top floors of the business associations and automobile manufacturers. "We were convinced that we were in solid accord with the rest of the industry," explains Bischoff disappointedly. In truth, however, there never really was any accord. "We were left in the lurch by the rest of the industry," complains the Daimler-Benz aviation and aerospace director. "Those who shouted loudest were the first to cave in."

Schrempp recalls these events with both frustration and anger in equal measure. He becomes enraged when he reflects on the commitments made by his management colleagues in other companies, for they proceeded to drop out of formation one after another.

Sports car manufacturer Porsche reached its decision at the start of October: The chairman of the board, Wendelin Wiedking, chose to adhere to the previous system of full sick pay. At the time Porsche was introducing a new product, the Boxter, and had no intention of risking strike action.

The head of Bosch excused himself because he was in the United States.

BMW's chairman, Bernd Pischetsrieder, who saw the concept of continued payments during sickness in the same light as Schrempp, was on a trip to South Africa at the crucial moment. He was prevented from making a decision, as he had given one of his directors full responsibility during his absence.

But the above-named managers were not exceptions. Privately, Manfred Bischoff makes no bones about whom he accuses of lacking steadfastness and being spineless. Schrempp does not wish to commit himself and speaks vaguely of the "vacillation of certain chairmen" that he found "dreadful."

Schrempp already felt that Heinrich von Pierer, the head of Siemens, had let him down during the discussion on demands for increased shareholder value, during which his indifferent attitude had noticeably annoyed Schrempp. No support had been forthcoming in the debate on sick pay, either. When the moment of truth arrived, Pierer took shelter in weak excuses about how the company's data-processing system would not be able to cope, says a Daimler board member who denounces the head of Siemens for abandoning them and categorizes this betrayal as nothing less than a broken promise.

Schrempp's retrospective explanation is witness to the deep wounds that this conflict inflicted on him: "We argued on the basis of law and it ended up being one legal argument against another."

At Daimler, too, there is no shortage of explanations for this knockout blow. Schrempp admits that "we miscalculated on this wage business," saying that the company was "not interested enough in the psychological background, only with the facts." Above all, however, the Group's head now acknowledges that "we did not carry out any historical research." By this he meant the fact that none of the board members was aware of the great significance of the union movement's victory in the historic fight for full sick-leave pay forty years earlier. "All in all we were badly prepared," says Schrempp, "and we did not argue convincingly enough." He adds also that "we didn't want to do any harm to those who were ill."

At the time, the decision to reverse policy was difficult, but there was no alternative (though when Schrempp made the announcement, two weeks after the fateful board meeting at which they voted to reduce sick pay, it almost sounded as if he regretted the step). He said that he and Gentz had received a mandate from the board authorizing them "to prevent any damage to the company," and he admitted that "we simply failed to get our ideas across properly."

Even if "in the end the company . . . sustained no material harm" as a result of the board vote and the subsequent reversal, Schrempp has to admit that on this one he lost. The media acclaimed Karl Feuerstein as the hero of the battle for full sick-leave pay—and poured their malice out over the man who had put himself at the forefront of those fighting for wage reductions.

The ultimate cost was that more than four thousand Mercedes vehicles were not produced because of the job action, sales were reduced by about $111 million, and the chairman's reputation was trampled. In addition, the Daimler boss underwent the painful experience of being betrayed when he wanted to be able to rely on his fellow fighters in the ranks of the automotive industry.

What avenues remain open to Schrempp to express his displeasure about how the other captains of industry buckled? He knows how to deal with similar situations that take place within the company. "I believe in one basic principle, and there are no exceptions to it, either: If an employee blunders, I will protect them from external interference as long as they have done nothing illegal." But Schrempp says this "does not of course mean that they will not be given a rap across the knuckles within the company." This approach is of only limited use outside the company. Schrempp must wait patiently until he is in a position to give his supposed friends in the automotive industry "a rap across the knuckles."

Standing Ovations

But our skilled workers must not be forced into packing
shopping bags at supermarkets after their early shift
so that they can support their families.
Jürgen E. Schrempp on November 14, 1996

In November 1996 the verbal battles over Schrempp's demands for 12 percent yields from all divisions, with shareholder value having top priority, had reached their peak.

It was only one month since IG Metall had used the issue of sick pay to serve Daimler's chairman the worst defeat he had yet suffered during his period in office. It was now a matter of regaining lost ground with the unionists and the Social Democrats and indicating to them that the Schrempp who was obviously on the side of the shareholders was also on the side of the workers.

This was undoubtedly a hard task. But Jürgen Schrempp really is a master of communication, and in situations where others would capitulate, he knows exactly which words to direct at which people. And, like no other, he can even win over an extremely critical public by using skilled rhetoric and deliberate gestures. This was what he would rely on when faced with the enmity of those who had workers' and society's interests as their priority.

"We are facing a critical threshold, a threshold behind which lies a country very different from the one we are familiar with," exhorted Schrempp in a speech at the Friedrich Ebert Foundation, which was closely aligned with Germany's Social Democratic Party. He referred to Germany's history in order to illustrate the full extent of the possible outcome and implored democrats in all of society's institutions—companies, unions, and the general public—to stand united in order to prevent this threshold from being crossed.

All had to "seek out solutions together, regardless of what responsibility we bear." Schrempp's socially responsible remarks climaxed with the promise "to breathe new life into the heavily burdened term *solidarity* using all the resources at the disposal of Daimler-Benz."

With such acknowledgments of the primary importance to the country of social policy and the secondary position of a profit orientation, Schrempp's calls appealed to the sympathies of the audience at the Friedrich Ebert Foundation. The applause was appreciatively long.

Had Jürgen Schrempp changed sides? Had the Rambo-like hardliner metamorphosed into a hero of the people? Many of the listeners that day may have thought so. After all, they were not present when the chairman of Daimler-Benz spoke before business representatives in Vienna less than two weeks later, talking about the challenges and opportunities of the globalization of business.

The audience in Austria's capital consisted of leading personalities on the Viennese stock exchange. With the fall of the Iron Curtain, this city had begun to regain its position as the geographical, political, and commercial center of Europe and had long since revived its reputation as an excellent location for those seeking to do business with Eastern Europe.

Schrempp's main theme was the expansion of the European Union eastward, which had to be seen against the background of "the global reapportionment of commercial achievement potential." Logically, no international consumer in any market was prepared to pay a higher price for a product just because it was made in Germany, and no customer was going to buy a car because it had a sticker saying that the manufacturer guaranteed its workers full sick pay. "That would not be a sales argument," said Schrempp, resuming his usual confrontational manner. "It is only the price of the product that counts, its efficiency, its quality, and its progressiveness." With this acknowledgment of the primary importance of a profit orientation and the secondary position of social aspects, Schrempp's calls appealed to the outlook of the bankers and speculators at Vienna's stock exchange. The applause was appreciatively long.

Had Jürgen Schrempp changed sides? Had the hero of the people mutated back into Rambo? Many of the listeners in Vienna may have thought so. They were, after all, not present when the chairman of Daimler-Benz spoke before the Social Democratic Party's central works councils a few days later, on aspects of globalization and full sick-leave pay.

The audience, consisting mostly of members and guests of party works groups at Mercedes-Benz AG, heard Schrempp admit that mistakes were made in the board's handling of the sick-pay matter and that it was wrong to reduce it to a matter of employment law. Schrempp frankly conceded that "I learned here that the many experts on employment law have many different opinions." For this reason all interested parties should "sit at a table together to find a solution that they could all accept and that, above all, improves our competitive position."

With this nod toward the primary importance of social aspects and the secondary position of a profit orientation, Schrempp's calls appealed to the Social Democrat audience, and the applause was again strong.

Had Jürgen Schrempp changed sides? Had the tough capitalist shape-shifted into a man with a social conscience? Many of the listeners in Stuttgart may have thought so. They were, after all, not present a few days later when the chairman of Daimler-Benz . . .

Examples could be rolled out at will. In reality the man on the eleventh floor of the Möhringen headquarters was in no way prepared to shift by even an inch from his views, of whose validity he was totally convinced. Those who know Schrempp know, too, that the Rambo-like manager and the social progressive are one and the same person. His varying perspectives were not a sign of a split personality but rather the expression of productive poles that augment each other.

If one of these two poles were absent, Schrempp would not be Schrempp and, critically, this man would never have become Lord of the Stars. Daimler's chairman had taken the expedient path during countless discussions: seeking consensus where it was advantageous for him or taking the harder, antagonistic course where this provided more progress—for the good of the Group and for his own good.

Without further ado Schrempp pulls out the small pad of paper that he usually carries around with him and draws what he describes as a "magic triangle." He scribbles the words *employee, customer,* and *shareholder* at each corner in his typically quick way of writing. The task of the company is to satisfy the needs of these three groups equally. "I have to overstress the justifiable interests of the shareholders because they were not taken into account before," explains Schrempp. In the next breath he adds, "Don't forget the other two pillars—the workers and the customers." Someone who is in the room asks, "Is he getting soft?"

"I had a crucial experience," says Schrempp, referring to his speech before the Social Democratic works councils in December 1996. The chairman frankly admits that "I was unsure about this meeting," and not without reason. During the months before the meeting Schrempp had acquired an image as a ruthless capitalist, and he did not know how five hundred Social Democrats gathered in the hall would treat the "Rambo of the nation."

Finally, however, he leans back in the sofa in his office and concludes: "The questions were very critical and more open than I am used to from my directors." But "they listened to me" and, what he finds more important, "the long applause gave me strength."

Even Schrempp must first learn to win over his supposed enemies. He finishes on a self-critical note: "I need more contact with people."

Nobody could accuse him of having made any false statements. In no speech before gatherings with representatives of employees, industry, or bankers did Jürgen Schrempp say anything that was untrue. It would be fairer to say that by leaving out certain factual elements the great communicator simply gave a one-sided presentation. He was almost always able to captivate listeners with a totally tailor-made talk.

In South Africa, Schrempp had already developed the ability to aim a certain speech at a certain audience. "I will never forget the first meeting," recalls Gerd Andreas, the automobile manager from Cape Town. At a dealer meeting in Durban, now twenty-five years ago, Shenker, then still chairman of MBSA, spoke first, followed by Schrempp. But only "Jürgen spoke so convincingly about technology that he received a standing ovation from the dealers"—which was more than his boss, Morris Shenker, got.

A Chancellor for All Cars

I am the Chancellor of all German automobiles.
Gerhard Schröder

With Social Democrat Gerhard Schröder leading the new German government, Schrempp has a good friend in high places. The two speak informally and comfortably with each other and can look back on many years of intensive and successful cooperation that began while the politician was still in charge of the state of Lower Saxony and the manager was still chairman of Dasa in Bavaria.

When the Christian Democrat state government in Baden-Württemberg was politically unable to force through permission to build a new Mercedes test track in Boxberg at the beginning of the nineties, Schrempp's friend came to the rescue. The high-speed course was built in Lower Saxony's Papenburg with the agreement of the Social Democrat/Green coalition under state leader Schröder. Thank you very much, Gerhard.

In order to return the favor, Dasa chairman Schrempp agreed to his political partner's request to spare Dasa's aviation works, located in Schröder's constituency, from threatened closure. The Social Democratic state leader made a deal with Schrempp, and Schröder founded a holding company to rescue the plant, and saved more than a thousand jobs. Thank you very much, Jürgen.

Schröder, for his part, helpfully supported the chairman of the defense giant through one of his worst crises. Dasa was mostly racking up heavy losses at the start of the nineties and urgently needed new defense contracts from Germany's Ministry of Defense. While Jürgen Schrempp was fighting Social Democrats, the Greens, and the large majority of the German people in 1993 to try to obtain billions in state subsidies to build Dasa's interceptor, the Eurofighter, he had full backing from his friend from Lower Saxony.

Schröder even risked jeopardizing his hold on the state government for Schrempp, when—against his own party's policy decision—he "trampled" the peace principles of Lower Saxony's governing coalition, in the words of the Green politician Andrea Hoops. Instead Schröder openly demanded (as Schrempp would have wanted) "an unprejudiced discussion on the Eurofighter." After all, Dasa's chairman was "a capable and reliable business partner," Schröder said in praise of him a year later. Thank you very much, Gerhard.

And so that everything continued to run on its ordered course, the Social Democratic party's coffers received financial contributions from the Group's well-filled gift account for many years. In 1993 alone, the year of Schröder's support for the Eurofighter, Daimler-Benz transferred nearly $213,000 to the SPD's account. And in the years that followed the party was also generously supported—just like their opponents, the Christian Democrats/CSU (Christian Social Union). Thank you very much, Jürgen.

At the end of 1998, however, the friendship between these two car enthusiasts had become strained. Gerhard Schröder broke with the time-honored tradition dictating that the German chancellor's official car come from the house of Mercedes-Benz. In answer to the question of whether he considered Chancellor Schröder's selection of a model from Volkswagen's subsidiary Audi as a challenge, Schrempp reacted defiantly, saying that he was not surprised that Schröder drove a car of this make in view of "his many years of activity on VW's supervisory board." "But we would endeavor," continued DaimlerChrysler's chairman, "after a while, to restore his enthusiasm for the best car in the world—a Mercedes."

In the spring of 1999, Schrempp finally achieved an overwhelming success in the get-the-chancellor-a-Mercedes project: At the opening of Volkswagen's new Berlin branch, Gerhard Schröder, contrary to his expectations, was chauffeured away from the house of Volkswagen by a reception committee from Germany's Federal Office of Criminal

Investigation in a Mercedes. Schröder, however, knew how to reestablish his authority. Shortly before arriving, the self-proclaimed "chancellor of all German automobiles" stepped out of the Schremppmobile and reached his destination on foot.

Profit Without a Stutter

If Daimler's Chairman elect were asked about his vision he would also, therefore, have an answer ready: 'profit, profit, profit'.
Der Spiegel

I deliberately exaggerated shareholder value.
Jürgen E. Schrempp

The outcome is an ongoing dedication to creating shareholder value as an objective at Daimler-Benz, now and in the future.
Jürgen E. Schrempp, before the Economic Club of Detroit

It sometimes took an astonishingly long time before this otherwise experienced speaker realized the impact of his use of certain words, particularly in his homeland. Jürgen Schrempp, in his function as chairman of Dasa, forced through a radical policy of layoffs called the "dollar-low rescue program" (known by its nickname, "Dolores"). In the beginning the aim was to reduce the number of Dasa employees from 86,000 in 1993 to less than 70,000 in 1996. In 1995 Schrempp, as Daimler chairman, decide to fire an additional 8,800 employees and closed three plants. Schrempp became one of the most hated managers in Germany at that time. Thousands of workers protested, but they were able to achieve only a cosmetic reduction in individual rationalization measures.

Schrempp, however, eventually recognized that part of the negative impact of his concept was due to semantic aspects, and he immediately renamed the painful program. After all, who could have anything against a "competitiveness initiative" in a free-market economy with competing companies and innovative products?

"A specter was again haunting Europe—and not the specter of Communism this time," stresses Schrempp. He's not referring to himself, either. This time he is talking about a term "that came from the Anglo-Saxon language" that was spreading "anxiety and agitation": *shareholder value*. When Schrempp realized that Daimler-Benz was perceived as being the trigger for this public debate, he decided he would no longer use such an "ideologically loaded" term in public in Germany, as he did not wish to "invite further misunderstandings." In his opinion, shareholder value means "nothing more than that the company management must aim to increase the value of the company." In Daimler-speak he now calls this process *increasing company value* or *value-oriented management*.

These are "exactly the same but in a form adapted to the German language." Furthermore, the phrase *increasing company value* is ideally suited "for avoiding misunderstandings" and "not overtaxing [people's] foreign-language abilities."

Schrempp's words are well chosen. In no way is he saying that he will never use the term *shareholder value* again. Rather, he chooses an alternative "when I speak German." He was thus saying that when he jets over to the United States as a head of DaimlerChrysler, he will continue to speak of shareholder value.

Just imagine, says Schrempp privately, "I am in situation X and detect a weakness. . . . If I want to improve it I must convince other people of my opinion." This leads to his having to "depict a situation more drastically than is really the case."

When he rose to the top of Daimler-Benz AG in 1995, the company was posting heavy losses. In such situations Schrempp did not shy away from denouncing the deplorable state of affairs at the highest decision-making levels. "I called up ten executives in 1995. Six of them," according to his sobering assessment, "did not know the value of the shares, and three [said they did but] quoted me a wrong figure."

He instigated the debate on the importance of shareholder value as a top priority, and he didn't have to wait long for the reaction of the media: "There was this *Der Spiegel* article with Schrempp demanding profit, profit, profit. I had been categorized."

But that categorization did not carry over to the United States, and Schrempp noted in a January 1997 speech to Detroit's Economic Club that "the world is quickly [developing into] a single large market . . . where buying and selling, investing and producing" were carried out. His hymn in praise of globalization peaked with the promise that Daimler-Benz had set shareholder value as a lasting aim—now and in the future. This was followed by the verses on reducing the number of divisions from thirty-five to twenty-three, obtaining a minimum yield on invested capital of at least 12 percent in the subsequent year at the latest, and pursuing the disposal of Fokker and Dornier that this would necessitate.

In the end, however, it is all the same. Regardless of whether he employs German or English vocabulary, the ideas behind the words are absolutely identical. And Schrempp is not particularly worried about criticism in Germany: "I am instantly understood when I go abroad." He knows his international reputation is an excellent one.

At the start of the nineties "we took too little notice of the shareholders' interests," he says today. In the meantime, however, "the central works councils have also understood that shareholder value meant increasing the value of the company." Naturally, this was "no short-term process," but even this late recognition by his former opponents still pleases him.

And Schrempp has long since regained his sense of humor. Thus today he can toss off a wry comment on the media and public criticism of his supposed demands for "profit, profit, profit," the comment that hurt him so badly at the time. "One profit was sufficient," opines the Daimler boss, and adds: "I didn't stutter."

Capitalist on a Collision Course

Jürgen Schrempp is clever enough to know that we are on
a course leading to social and ecological confrontation.
Professor Ernst Ulrich von Weizsäcker,
President of the Wuppertal Institute

When asked what he sees as the consequences of globalization and whether he predicts a worsening in the ecological crisis, Erich Ulrich von Weizsäcker, former head of the Wuppertal Institute (an international scientific foundation with a stellar reputation in environmental protection), does not mince words. "Globalization is an extremely dangerous phenomenon for the environment—particularly when managers in Thailand, South Africa, or Germany see themselves as being under pressure to increase capital gain—which almost rules out long-term ecological considerations." Therefore, "one cannot be surprised that the overexploitation of natural resources has accelerated since 1990." Daimler-Benz is "simultaneously one of the largest suppliers to exploitative industries such as the mining industry," in that it supplies the fleets of trucks.

And if one asks him about the role in this process played by the chairman of Germany's largest company, Weizsäcker's position is no less clear: "I would classify [Jürgen E. Schrempp] as the leading representative of the dangerous spirit of the times." In von Weizsäcker's view, Daimler's boss is "a pioneer among the globalizers in German industry—ahead even of Dormann (head of Hoescht), Pierer (head of Siemens), and Henkel (president of the National Association of German Industries)." And, according to von Weizsäcker, the term *globalization* is "closely connected to the term *shareholder value*," which "was first brought to public attention in connection with Jürgen Schrempp."

Comments such as these are hard to take for someone who is so passionately enthusiastic about the idea of globalization.

"I have no comment to make" is Schrempp's first reaction to the respected ecologist's criticisms. It is understandable that Daimler's chairman does not like such descriptions. Then Schrempp decides that he will answer the points raised after all. "Time will decide which view was correct," he says, qualifying his dismissive attitude.

No other large German company has "smashed firms" to this extent and so consistently destroyed jobs, according to complaints from the unions. While in 1991 almost 380,000 people were still employed by Daimler-Benz AG, by 1996 the number of workers had been reduced to just 290,000.

It would be just as wrong to make Jürgen Schrempp solely responsible for this as it would be to exonerate him of all responsibility. Edzard

Reuter led the activities of this public limited company formally until May 1995 and de facto until the summer of the previous year. If one were well disposed toward Reuter, one would commend him for the fact that during his time as chairman of the board of management, from 1987 to 1995, there was no overall reduction in the number of jobs but rather an expansion followed by a contraction. If one were well disposed toward Schrempp, one could ascribe to him the first increase in the number of employees after 1996.

It is, however, a fact that in the first half of the nineties alone, the board was responsible for decisions that resulted in the actual loss of forty-five thousand jobs. Eckhard Cordes sees this differently insofar as Daimler-Benz hasn't "fired a single person but dismantled jobs under market pressure." Furthermore, "there was no active dismissal." Rather, "our employees left us by mutual agreement."

Competition on the global mobility market was plainly becoming tougher. The merger of Daimler and Chrysler put Jürgen Schrempp at the top of the list of merger makers. At the same time, however, he suppressed other questions: Were those analyses by leading research institutes correct when they said that there were already too many, and not too few, automobiles being produced in Europe and the United States? And how did these forecasts match the demands of Daimler's chairman for a massive increase in vehicle production in the coming years? And what climatic consequences, Mr. Schrempp, were to be feared in view of the fact that Daimler's fleet of passenger cars scored extremely low in terms of fuel efficiency?

Weizsäcker, a Social Democrat who now has a seat in Germany's lower house of parliament, also links the issue of the global division of labor with the Daimler chairman's name: "The present model of the international division of labor in world business is that of Jürgen Schrempp." Whose comment on this criticism was short and to the point: "Correct. He really is right about that."

Then Schrempp refers to the central reason, from his point of view, for locating the production plant for M-class off-road vehicles in Alabama: "We haven't built a single plant at a particular location because of the international division of labor, not even the one in Tuscaloosa";

rather, "we go where the market is," and, as everybody knows, this is not restricted to Germany. This argument may well apply for the Mercedes works in Alabama. The market for fun cars is certainly much larger there than in Europe, where space is limited, environmental regulations are more restrictive, and the number of people interested in this type of vehicle is smaller. However, with this line of argument Jürgen Schrempp fails to mention that Jim Folsom Jr., governor of Alabama at the time, put together the most attractive package out of about two hundred other offers, including low taxes, an infrastructure built for nothing, and grants for training courses.

In the meantime, Daimler's chairman of the board has become so influential that his approval of any particular location on earth attracts investment amounting to billions, causes factories as large as towns to rise out of the ground, and provides work for thousands. With an eye toward this enormous power, politicians all over the world woo the Lord of the Stars. The presidential red carpet is rolled out when he jets abroad—and social and environmental issues are swept under it, out of sight.

Daimler's chairman answers his critics with the claim that not a single plant has been "located abroad for reasons of cost." He skillfully adopts the classic unionist argument of international solidarity as his own. "Social responsibility was not only relevant in Germany," according to Schrempp, who claims that his business responsibility extends equally to workers in South Africa's East London and Alabama in the United States. "In part [our responsibility] was our activities in other countries with extremely high unemployment and no social safety net," he stresses, presumably not completely without ulterior motive. Schrempp's classic rhetoric on jobs usually leads to his raising the argument that "three jobs abroad secure one in Germany."

At least when they eventually met the two prophets were communicative, not confrontational. Schrempp praised the then-head of the Wuppertal Institute as "a good thinker," and anyway there was nothing wrong "in having different opinions." He who had been so highly honored replied, "It would be idiotic of me to nail these criticisms to the man Jürgen Schrempp," as he has simply "achieved the best for his company and the business location of Germany under conditions that were not his choice." Besides, he added, managers are "mostly selected on the basis of whether they, like Jürgen Schrempp, are particularly good at raising the yield gained from capital."

Duel of the Nameless

Reuter's statements on shareholder value and capitalism are dangerous in their one-sidedness because they are misleading.

Jürgen E. Schrempp, Edzard Reuter's successor

Jürgen Schrempp does not merely talk about globalization, he practices it—as an actor at the forefront, and not without opposition even from his predecessor.

Those who see the conflict between the present and the former chairmen solely from the point of view that Reuter stood in the way of Schrempp's career ignore the fundamental differences regarding content that set the rivals apart—then as now. And almost none of Reuter's other criticisms hits Schrempp harder than his warnings against unrestrained capitalism and purely profit-oriented management.

When Edzard Reuter talks about today's capitalism and the present leadership elite, it is as if the Zeus of the eighties were striking the management generation of the nineties with all the concentrated power of his bolts of lightning. Reuter refers to one of his speeches in which he explained that so-called experts maintained that the sole obligation of companies in market economies is "to increase yield or, more popularly expressed, profit." These experts are as little concerned with the interests of the community as with the interests of people who have jobs at the companies. The result was "the pure theory of the market economy that rejects integration of any type, even of a social nature, as inadmissible dilution and even as the work of the devil." Reuter says that he has "never agreed" with this theory.

The intellectual prophet still sees ways out of a development that could lead humanity into a catastrophe of undreamed-of proportions. If the process of globalization "is not to end in chaos," then "control on the basis of democratically defined objectives" must take place. But the democratic system is daily threatened with the loss of its legitimization as a result of "the unbridled predominance of economic affairs."

The chairman of the board of Daimler-Benz AG has a completely different view. Reuter's statements on shareholder value and capitalism, he says, are "dangerous in their one-sidedness." It suddenly becomes clear to even the most remote of observers how deeply his predecessor's words hurt the current head of Daimler.

Of course capitalism has its limits; of course it must not be allowed to run riot. "The limit of capitalism has been reached," stresses Schrempp energetically, "when responsibility ceases to be considered in the long term." But instead of defining responsibility more precisely and ensuring that it is strictly observed, the head of Daimler criticizes short-term thinking. "If one is only interested in quick money," Schrempp concludes, "that is survival-threatening." In this sense Schrempp describes himself as "an opponent of short-term maximization of profits as practiced, to some extent, in the United States."

Anyone who expected more than this will be disappointed. There is nothing about the negative effects of globalization, about the destruction of small businesses by large ones, about millions of people being forced into poverty, about mass unemployment or ecological destruction. Instead Schrempp proffers shareholder-oriented thinking: "Short-term optimization is anti-shareholder." It means that research and development, as well as the encouragement of training, decrease in importance. "That would mean," according to Schrempp, "devouring the company's seed potatoes."

There is more behind such statements than the dry analysis of an angry chairman of the board. Jürgen Schrempp is taking aim at Edzard Reuter's now-lost vision of creating a profitable integrated technological concern. Ever since this plan was derailed, correcting the share value has become the top priority for Schrempp—who was at that time actively involved in the portfolio decisions made. His enthusiasm for the short-term thinking of political decision makers was limited: "It would be hopeless to run a company like a minister of finance or business would."

Nonetheless, Schrempp sees light at the end of the tunnel. "Luckily, the debate in Germany is becoming more rational again," says the Daimler boss happily.

So much for the logical arguments that, once one considers whom these apparently objective statements were originally aimed at, quickly are seen to be personal. "At first glance it seems as if [we are in a time] of ice-cold calculators, of young analysts untroubled by emotions, of the power people so beloved by the media," ruminates a thoughtful Reuter. He names no names, and certainly not *the* name.

Edzard Reuter's aim is the restoration of the worldwide primacy of politics and on no account its unconditional surrender to the markets. For Reuter, the primacy of politics remains "inalienable," if "the basic premises of a democratic state are not to be brought into question." The community must have precedence over the egotistical interests of the individual. "This, and nothing else, is the duty of a management elite deserving of such a name." This comment is directed toward a limited number of people. There is only one he means, however, when he emphasizes that "those who are responsible for business undertakings count among them."

Neither names names. Reuter does not when he means Schrempp, and Schrempp does not when he means Reuter. And despite this, the target of each individual attack is obvious. The rift between the former and the present chairmen of the board of Daimler-Benz AG is so wide that it will never be bridged. Apart from the factual disagreements on the fundamental issues of globalization and the ethics of business policy, the two top managers are also united by a mutual lack of understanding that cannot easily be overcome.

Jürgen Schrempp created his incomparably better starting position at Edzard Reuter's expense. Even if Reuter were to quote the virtues of age another thousand times, he would never be able to forgive his former protégé. And Schrempp could never repair the damage done, even if he were to attempt to—which is extremely unlikely. The split appears irreversible. There will be no winners in this duel of the nameless.

CHAPTER 8

Schrempp Superstar

There is nothing to restructure here.
Jürgen Schrempp on Euclid Inc.

Schrempp was about to be thrown in the deep end again. Chairman of the board Gerhard Prinz personally phoned Schrempp in his office in Pretoria and told the technical director of United Car and Diesel Distribution to get on that night's flight to Germany. In Prinz's office less than eighteen hours later, the chairman presented Schrempp with some highly confidential facts. Schrempp listened, in his impatient way, to a company history that became more interesting as Daimler's chairman shed more light on the situation. In the end he understood why Gerhard Prinz had summoned him: The next test for Schrempp awaited him in Cleveland, Ohio.

Mercedes-Benz at this time was already producing a wide range of commercial vehicles. During the seventies and early eighties the company achieved higher sales in this sector than with their passenger cars. But in the early eighties a negative trend developed, and the importance of the commercial vehicle sector was waning. In 1983 the board decided that some strategic decisions were called for, and began to look for a U.S. company to buy as part of its diversification policy. This not only would provide wider market access but also was intended to increase experience with new management techniques.

Gerhard Prinz, the board member responsible for materials under chairman of the board Joachim Zahn, had had the idea that Daimler-Benz should invest in producing the heavy mining vehicles required for transporting coal, oil-bearing shale, and earth in huge mines. He conducted the negotiations with Euclid Inc., the Cleveland-based producer of heavy transporters with a working load of up to 170 metric tons, and made the purchase in 1977. Much was expected of Euclid's business prospects, for this was "our first really new acquisition outside Germany," according to Edzard Reuter, who noted that until this point Daimler-Benz had "only taken over partners that manufactured or sold our traditional products."

Global developments, however, had long since spoiled the plans of Daimler's board. Since the second oil crisis in 1981, the United States had been having second thoughts about using oil-bearing shale resources, whose exploitation was simply too expensive. In addition, the world's oil deposits had turned out to be of far greater magnitude than had previously been assumed.

And to make matters worse from the point of view of those affected at Euclid, the OPEC countries had meanwhile come to an agreement over the price of oil.

In the end, it had all backfired on Daimler-Benz. The hoped-for boom in heavy vehicles had failed to materialize, Euclid was in a deep crisis—and Schrempp was supposed to save what was already past saving.

"He should show what he is capable of on this." These were the words that Daimler chairman Joachim Zahn supposedly said to a group of confidants regarding the abilities of his successor, Prinz. A solution to the Euclid problem had been right at the top of Prinz's list of priorities since then. Schrempp confirms this impression, saying that the restructuring of the Daimler subsidiary was of "tremendous importance" to Prinz. In the end, however, everything went very differently than had been hoped.

A few years earlier he would have jumped for joy at receiving such an opportunity. He had always wanted to work abroad, especially in the United States. Now, however, the offer had come at a time when, after eight years in South Africa, he and his family had gotten used to living there and he had worked his way up from national service manager to technical director on the board of management. And Schrempp was facing a particularly difficult task.

So the Daimler chairman's offer had only limited appeal. The economic conditions that would complicate his new responsibilities were already in place; the chances of a career advance if he failed, which was not unlikely, were extremely questionable. Schrempp himself was well aware that he was absolutely clueless about the country and its people. Besides, the thirty-seven-year-old felt that he was simply too young and inexperienced for the task facing him.

"Should we really go to Cleveland?" Schrempp was unsure and was trying to put off the decision. In his indecision he called on his wife, who had stood by him through all the years and also given him business advice. Renate Schrempp first had to check in an atlas to find out exactly where Cleveland was, but when she did, her reaction was emotional: "Immediately! With the lake so near, it would look like Zurich."

To be on the safe side, she phoned the American ambassador and asked about conditions in Cleveland. The answer was rather disillusioning. "That's where the rivers burn" was the gloomy description of the pollution in the area, and downtown Cleveland was supposed to be a "smokestack"—this was the less-than-attractive prospect facing them if they decided to move there.

When he weighed the pros and cons, there seemed to be more against the move to Cleveland from Schrempp's point of view, so he sought support from Peter Emil Rupp, head of Euclid. "Tell Prinz that I am not yet ready," Schrempp requested. "I had no idea about the United States. I had no idea about off-highway trucks," Schrempp says of himself back then. "I felt I was still too young."

Rupp—who, considering the strained business situation, was not uninterested in moving on—promised to help but did the very opposite: "Schrempp was exactly the right man," he told the chairman of the Daimler-Benz board, referring to his foreign experience and his talent for dealing with problems head-on.

Gerhard Prinz made no secret of why he specifically selected Schrempp to knock Daimler's ailing subsidiary back into shape. He had "the impression that Jürgen Schrempp, with his analytical style, would approach the problem with his mixture of insolence, directness, and honesty and simultaneously make suggestions about how to solve it." Besides, Schrempp came from the company's commercial vehicles division and had been recommended in South Africa as an effective troubleshooter—at least, this was the argument Gerhard Liener used to draw Prinz's attention to the UCDD board member.

While Rupp, as expected, moved to the top of Freightliner, a producer of heavy commercial trucks that Prinz had bought for Daimler-Benz in March 1981, Schrempp—rather surprisingly—put his reservations behind him and submitted to the chairman's pressure. Schrempp left UCDD's board on September 15, 1982, moved to Ohio with his family, and took up his new position as president of Euclid in the following month. His task was to restructure the firm.

By 1981 Euclid was selling few of its enormous trucks. The U.S. economy was in desperately poor shape, prices were tumbling, and the market had collapsed to a quarter of its former size. At the same time the dollar was reaching new heights, increasing the prices of vehicles produced in the United States and drastically affecting export, which accounted for two-thirds of Euclid's total trade. Nevertheless, Euclid had still managed to produce 604 dump trucks in 1981. Compare this with the fewer than 400 that were made in 1983 under Schrempp's chairmanship. "A lot of markets have dried up," moaned a visibly unnerved Gerhard Prinz.

These "really powerful vehicles" have a "particularly sensual attraction" that "could sometimes affect the objectivity of one's judgment." This was how Edzard Reuter described his first impression of the enormous heavy transporters, which he felt were "wonderful." With each month that passed, it became increasingly evident that Daimler's chairman had made an unwise investment. In fierce competition, the Japanese supplier Komatsu Ltd. proved overwhelming.

The attempt to redevelop the vehicles also failed, as did the reorganization of production procedures. "Euclid had almost more loss than total sales," states Christoph Köpke soberly. Euclid was forced to reduce its workforce, but there were still 1,151 employees in Schrempp's first year, producing sales of just $188 million.

Jürgen Schrempp labored hard for more than a year—to no avail. Indeed, the company's rapid decline continued, with losses increasing. But when Schrempp finally realized that the "restructuring project" was doomed to fail, he slammed on the brakes in his own inimitable style.

There was nothing to restructure, said Schrempp. He was not one to beat around the bush, and his harsh conclusion was that the company had to be disposed of. The attempt to save the U.S. subsidiary was absolutely hopeless, Schrempp told his boss during a private meeting. He was well aware that Daimler-Benz had never before disposed of a part of the company. And he was taking a considerable risk—with this statement he was sawing off the branch on which he himself was sitting. Daimler without Euclid was also Schrempp without a job.

An increasingly worried Daimler boss delved deeper and posed new questions in the hope of finding a solution: What did the actual figures suggest? Was there any way out? What alternatives were there? Despite this, he chose not to make any remarks critical of the man on whom he had pinned his hopes. And after this meeting he had no alternative save admitting to Schrempp that he was right—but that "we will sell in Daimler-Benz style." And by this he meant protecting the employees.

Only three people were involved in the secret negotiations: Gerhard Liener, Gerhard Prinz, and Jürgen Schrempp. Daimler's chairman phoned the chairman of Euclid frequently to hear the latest news. Schrempp conducted talks with three different business partners simultaneously. He wanted to be in a position to present the draft contract to Daimler's head by the end of October 1983. But it never came to that.

It was a complete surprise when Gerhard Prinz died on the afternoon of October 29, 1983. Schrempp, who during the process of Euclid's sale had called him at least daily and regularly jetted to and fro across the Atlantic, was "really upset" by the news. He knew that the coming months would be equally hard for both the Daimler board and himself. A new chairman had to be chosen and—of vital importance to Schrempp—the Euclid deal had to be brought to as successful a close as possible.

From then on Euclid's boss had to play his part without any assistance from his former superior. He had to rely on himself—and nonetheless managed to handle his sales negotiations with a great deal of skill.

When James D. Rinehart, head of the Clark Equipment Company, visited the Daimler-Benz headquarters, he was welcomed particularly cordially by Werner Breitschwerdt. Considering the disastrous situation, Daimler's team was delighted that any buyer at all could be found. And Rinehart was also pleased, for buying Euclid represented a logical extension of Clark's range of products, which included axles, gears, machine tools, and even components for conveyor systems. And at this reasonable price Euclid's dump trucks fit into his program very well.

In exchange, Daimler-Benz AG acquired a 5 percent capital holding in the U.S. machine tool company. This was worth $30 million—exactly the same amount that Daimler had spent buying Euclid seven years earlier. Against this, the two capital infusions of $40 million each (in 1982 and 1983) were irretrievably lost. And the actual settlement turned out to be so meager that Jürgen Schrempp did not, at first, dare announce the exact amount publicly.

While all this was going on, the business of Daimler's subsidiary Freightliner was flourishing. In 1983 production rose by a sensational 50 percent to twelve thousand trucks. The satisfactory level of incoming orders ensured capacity utilization and the jobs of 4,800 employees.

The passenger car business was also booming. More than seventy thousand luxury limousines were sold in the United States in 1983 alone, despite an average price of well over $30,000. Annual sales of a hundred thousand cars were even being forecast for the next six years. The U.S. importation company, Mercedes-Benz of North America, had meanwhile achieved sales of about $3.2 billion that year.

Contentment reigned all around when Rinehart and Breitschwerdt signed the final Euclid contract on January 5, 1984. Even the head of Euclid, who had made the best of a lousy situation, was satisfied, despite the comment in a German political magazine that the 5 percent share Daimler had acquired in Clark was "just a face-saving measure" and that the U.S. company had bought Euclid "for practically nothing." All in all, Schrempp and Daimler's financial director Reuter had accumulated a total loss amounting to around $53 million. The Euclid investment could confidently be considered a failure—but while this spelled the end of Euclid, it was not the end of Jürgen Schrempp.

Though his task had been clearly defined—turn Euclid around, make a profit—Schrempp had utterly failed to carry out this assignment.

But anyone who thinks that the German left the United States as a loser is making a big mistake.

It is typical of Schrempp that he emerges as a victor where others would be deemed to have failed. Euclid was just one example of a whole chain of partial successes and failures taking him from the losses in South Africa to the top of the whole Group, via the Dasa disaster and his involvement in the mistaken acquisition of Fokker. The key to his success lies less in the economic results of his work than in his unique knack of presenting his situation to others. He does it in such a way that they believe there is no other way to handle matters than his and that they would have made exactly the same decisions if they had been in his position. Schrempp is a world champion in advertising—above all for himself.

Listen to how other top executives describe his work at Euclid. Herbert Henzler, who until the fall of 1998 was chairman of the McKinsey Group's German branch, considers Schrempp the "troubleshooter" at Euclid.

Werner Breitschwerdt is fulsome in his praise of Schrempp's problem-solving capabilities and says that he sold Euclid "with extraordinary skill and in the process made a good deal," both forcing the buyer to "expand the scope of his offer" and, still more important from the point of view of Prinz's successor, allowing Daimler-Benz to get rid of a subsidiary "that stood apart from the rest of the company." Breitschwerdt still remembers that the closing of the Euclid contract, the first matter he had to deal with in his new position as chairman of the board, was exceptionally well prepared by Schrempp.

The reward for Schrempp's supposedly successful U.S. involvement followed immediately. On April 1, 1984, he was appointed acting member of the board of management of Daimler-Benz of South Africa.

From Werner Breitschwerdt's standpoint, Schrempp managed the difficult conditions in Pretoria and Cleveland very well: "I suggested that Schrempp should become director of the commercial vehicles division," recalls the former head of Daimler.

Edzard Reuter, in his memoirs, also praised Jürgen Schrempp's performance as chairman of Euclid very highly, though he added, "But nothing bore fruit. In response to a suggestion from Gerhard Prinz we were forced, in 1983, to decide to relinquish our purchase. With the sale of Euclid to the Clark company, the episode came to a final conclusion shortly after his death." Not a word about Schrempp's involvement in the sale.

Easy Times

The winners ... will be those who search out and participate in
the real growth industries and insist upon being number one
or number two in every business they are in. . .
Jack Welch, Chairman of General Electric

We are the number one in relevant passenger car business, we are
the number one for commercial vehicles over six tons, we are the
number one in the rail sector and the number two in aircraft.
Jürgen E. Schrempp, Chairman of Daimler-Benz

A magazine journalist wanted to interview Jack Welch Jr., head of General Electric (GE). "Do you think I should do this?" he asked Schrempp, his friend and confidant in Germany. The German approved, and the meeting with the reporter from *Headhunter* took place.

The bond that joins Jack Welch and Jürgen Schrempp is intense. "We have a very friendly relationship," says Schrempp. It is not surprising that they get on so well—they have a lot in common, including the rigorousness of their dealings and a certain hardness toward themselves and others.

"The employees would go through fire for Jack," says Schrempp. At least those who are still on the workforce, he should have added, for Welch, whose office is in Rockefeller Center in New York, cut the number of employees from the 412,000 who were working for the company when Welch first took up his position to only 222,000. Not even Schrempp has restructured that ruthlessly.

Some other things they share are their aims and their difficulties with the job of chairman. "We had to sell factories and fire people. That is painful for all of us and the worst part of my job," Welch says. "I feel really bad that people have to be fired. Livings depend on these jobs, whole families," Schrempp remarks.

"We gave the people decent settlements," emphasizes Welch, "and we didn't simply throw anyone out onto the street." And Schrempp? He says that while "reducing personnel is not our primary aim," unavoidable lay-offs "will, however, be dealt with fairly."

"If you want to create shareholder value, you need, above all, contented and motivated employees," stresses Welch. "We can only manufacture excellent products with motivated, well-trained and imaginative employees," Schrempp points out.

"We are rehiring people today," announces a proud Welch. An equally proud Schrempp commented in the spring of 1998, "I am sure that this year we will again create 2,000 to 4,000 new jobs in the Group."

General Electric is in the United States what Daimler-Benz is in Germany, from the tough rationalization programs to the lucrative stock options for employees and the huge increase in each company's value on the stock exchange. And the Jack Welch of Germany is Jürgen Schrempp. "I can see that there are dramatic changes taking place in Germany," the American says in praise of the measures undertaken by German company directors, and expressly names Daimler-Benz AG. Welch has the "courage to implement," the German says of the American's performance.

Welch's philosophy is as simple as it is radical. "We must be quicker, nimbler, and more competitive" is his recipe for success, and he sets his sights correspondingly high: GE must "be the number one or number two in every field of business in which we are active." When the forty-five-year-old Welch became GE's youngest CEO and chairman in April 1981, only the aircraft engine and plastic products divisions were globally oriented. Fourteen years later, in the same year that Schrempp was promoted to chairman of the board of Daimler-Benz, twelve of GE's divisions were ranked first or second in a worldwide comparison of their respective markets.

The German has no intention of being any less successful. "Every company must be viable, efficient, successful in the marketplace, innovative and profitable," he says. "We want to be the world's best in our areas of competence, and nothing less."

The actions that the two men have taken are in many ways as similar as their philosophies. Welch conducted a radical restructuring during the "hardware phase" of the eighties, and 350 areas of business were reduced to 13 high-tech and service divisions. Welch sold whole divisions for as much as $15 billion and bought or built new ones that better served his aims for up to $24 billion. Many observers were surprised that this resulted in increased productivity, sales, and profits, despite ever-decreasing numbers of employees.

GE's chairman could report total sales of $70 billion with a net profit amounting to $6.6 billion for fiscal year 1995. In the year when Schrempp was promoted from Dasa chairman to Daimler chairman, Daimler-Benz posted a record loss of $3 billion on sales of $54 billion.

However, by this time Schrempp had long since laid down the framework necessary for further development of Daimler's portfolio. Schrempp did not mince words when he demanded that the management and supervisory boards offload the companies he classified as "the main loss makers," namely, Fokker, Dasa, Unimog, and parts of AEG. In a parallel process, the board was to intensify its portfolio management. "We are concentrating on profitability," the fields of business, and "the strategic fit with the Group's core capabilities and core businesses"—that was the new line with which a slimmer, highly efficient company, with all its constituent parts finely attuned to each other, was to be created.

The German supervisory board members faced an American-style storm when Jürgen Schrempp made them aware of the current state of the portfolio process. At the November 8, 1995, meeting, Daimler's boss said, "We want to concentrate on the businesses we understand." Only activities "with long-term, sustainable profitability" would be continued, "and we will increase our international creation of value" by "conquering new markets" with products already in existence, by "product innovation in existing markets," and by the "development of new products" for emerging markets such as those of Southeast Asia.

Schrempp was aware of the worries of his central works councils, and he stressed that the reconstruction of the portfolio did not "necessarily" mean "the transfer of our jobs abroad." He explained that the

German market offered potential that needed to be utilized in the future. Furthermore, he promised to "involve the personnel representatives in the way that has, in the meantime, proved so effective."

Such nice words were all very well, but the reality was harsher. Inspection of the Group's structure resulted in a series of further disinvestment measures, leading to the sale of parts of the business and in some cases the laying off of thousands of workers. Apart from the well-known examples (from AEG Schneider Automatisierung—a subsidiary of AEG—to Regioprops and medical technology), plant and automation technology, energy distribution, and airplane engines (facing strong competition from BMW) were all on Schrempp's hit list.

Daimler-Benz in 1995 was undergoing radical change. The intermediate result of Schrempp's portfolio analysis was to highlight the strengths of the company. In the company's main area, mobility and associated services, it had the capacity and capability "to find integrated solutions" to all customer problems and market requirements for traffic and transport "by road, rail, or air." Only the "whole spectrum of traffic carriers" would permit supplementary system solutions to be offered "for urgent traffic problems," from traffic systems for cities to logistical conceptions for freight. Projects involving communications and computers, airport systems, or traffic assessment systems could be successfully implemented by means of interconnected networks. Daimler's chairman proudly announced to his supervisory board that "the spectrum of our activities in this area has achieved a breadth greater than that of any other company in the world."

However one evaluates the work of the man at the top of GE, Schrempp has learned a lot from him. Daimler's new portfolio bears the stamp of a radical restructurer in the mold of Welch. "Our future strategy is clear and unmistakable," the Group's chairman says, describing the route for Daimler in 2000 and beyond. In the coming millennium the company will fulfill the "comprehensive role of a mobility concern" in addition to having "a workable position in allied businesses." Daimler-Benz AG started making headway in this direction decades ago. But the Reuter era, with its policy of hesitant stalling and insincere calculated optimism, was over. What followed was the Schrempp era, characterized by consistent decision making and a considerable degree of ruthlessness.

Gone are the days of wild acquisitions beyond the scope of automobiles; Edzard Reuter's "integrated technological concern" has been buried.

"We limit our scope to what we are capable of," says Schrempp, sounding like the Welch fan he is, and proudly expounds on the fields of business in which Daimler is already the world's number one or number two. The past of the mobility concern lives on, supplemented and refreshed by the demands of a Daimler chairman who thinks and acts globally. The restructurer Schrempp is the globalizer Schrempp. And with him the cozy times of internationalization come to an abrupt end. Where the path leads is already evident: "Sometime we will look back on the nineties and say, 'Man, those were easy times,'" prophesies Jack Welch.

A Gamma World

Schrempp is really proud; he was hovering at least
one centimeter above the ground.
Participant at the extraordinary Supervisory
Board meeting of May 1998

The invitations had drifted in one week earlier, and nobody really knew what was actually going on. It was very unusual for the supervisory board to be rounded up for an extraordinary meeting—at least not without a clear agenda. One could have assumed that it would be about the suitability of Mercedes manager Jürgen Hubbert's idea of building a new passenger car works—and to upbraid him if necessary, for there had been astounded displeasure behind the scenes when, a few days before this special meeting, he had persisted with his idea of building a new "peak load factory" to reduce the considerable waiting periods for ordered vehicles.

Christoph Walther, the press office manager, to whom this had come as a total surprise, was not the only one who was dumbfounded. If there was going to be a new factory, the company's communications manager would have arranged a large-scale press conference to announce the happy news—as a rule, Schrempp would not pass up such an opportunity.

Hubbert's maverick idea had also caused considerable unrest among his colleagues on the board of management. "I was flabbergasted," said Heiner Tropitzsch, according to a supervisory board member, when delivering his personnel and social report at a meeting of the European central works council at headquarters in his function as labor relations director.

On May 6, 1998, at exactly 5:00 P.M., Hilmar Kopper welcomed the assembled board members to a remarkable meeting that, more than any before it, merits the description "sensational." Not even the extraordinary January meetings of 1996 and 1997, where the sale of Fokker and the new portfolio were discussed, could match this one.

The moment had arrived for Schrempp to drop his bombshell— Daimler-Benz AG was to merge with the Chrysler Corporation, the United States' third largest automobile concern.

The great victor derived intense pleasure from this success and was proud of himself and his team, who had been holding top-secret talks with their opposite numbers in the United States for more than three months. Just eight of them planned and argued, conceptualized and discussed: on one side Jürgen Schrempp, Eckhard Cordes, and Rüdiger Grube, and on the other Robert J. Eaton, financial director Gary C. Valade, and treasurer Thomas P. Capo. The investment bankers Alexander Dibelius (for Daimler) and Steve Bott (for Chrysler) had also joined them.

The deal proved how vital it had been for Schrempp to replace 60 percent of the staff in the company's strategy department in just one year. It had fallen to Grube to turn the new team into an elite squad. Eckhard Cordes and Schrempp had gone all out on the merger negotiations during the last few weeks.

"They were fantastic strategists," said one of the supervisory board members with respect, for without Cordes and Grube the megadeal would not have turned out so successfully.

After making the initial announcement, Schrempp leaned back contentedly as Hilmar Kopper handed the floor over to Eckhard Cordes, the board of management member responsible for direct industrial mergers and acquisitions. Cordes presented the basic structure of the new DaimlerChrysler construct to the still shocked-looking members of the supervisory board, highlighting the leadership duo of Schrempp and Eaton, the amalgamation of the boards of management and the supervisory boards, two headquarters—one in Stuttgart, the other in Auburn Hills.

Finally Jürgen Hubbert, who was to be responsible for Stuttgart's vehicles division on the future DaimlerChrysler board of management, had his say. The Mercedes man described the prospects of the merger of these two automobile firms with such rich traditions.

Second thoughts are not in Schrempp's nature. At most he would admit that the integration of the two corporate cultures would be a great challenge for the management. The Chrysler Corporation was considered the most patriotically led American automobile company and had even used the slogan "Americans buy American" in previous years. In the next sentence, however, Schrempp swept all doubts aside, saying, "I am sure that we will manage it." At the same time, as always, he was grinning extremely winningly.

One part of the deal made several members of the controlling committee look around disbelievingly. Daimler's board of management, pared down to ten members only two years earlier, would soon be swollen to almost double that number—seventeen. Would not flexibility suffer?

"That's almost two soccer teams," said one member to no one in particular. But Hilmar Kopper put a light gloss on the whole thing, joking that the board members would simply have to sit in two rows.

Hardly had news of this so-called elephant wedding gotten around than share values exploded on one side of the world and plummeted on the other. The value of American depository receipts rose on Wall Street by seven, to $109, thanks to Daimler and Chrysler; the price of Chrysler Corporation shares increased by six, to $47. Daimler shares also rapidly increased by 7.7 percent on the German stock index (DAX), and the trend continued. On the other hand, shares in Tokyo lost more than 2 percent.

DaimlerChrysler! The concept was astounding, yet when Jürgen Schrempp has made up his mind, it is best not to tangle with him. The question of the intercompany relationship was settled, and also that of who would have first position in the name: DaimlerChrysler, or there would be no merger. Schrempp had risked all, simply for the prestige of having the Daimler name first. And the risk had been great, for Robert Eaton was an experienced manager, no easy nut to crack. The counterproposal was Chrysler-Daimler-Benz—the Americans in front but the German compound name in full to make up for it—but Schrempp would not take second place. No way. Negotiations had continued late into the night on that issue, and if Chrysler's boss, Eaton, had not given in, then the megadeal would have collapsed simply because of the name. It really is best not to tangle with Schrempp.

Linked with the Chrysler Corporation, Daimler-Benz AG would immediately more than double its annual sales, to $154 billion. But those who know Jürgen Schrempp would not be too surprised that he soon publicly promised to double this again during the coming decade—by taking over other well-known automobile companies, of course.

Though the details of the deal were largely finalized and a date had been set for tying the knot, the Chrysler Corporation had by no means been Schrempp's preferred choice. If the ideas of Daimler's CEO and his negotiating partner Alex Trotman, from Ford, had been reconcilable, then the new global player would have been Ford-Daimler-Benz. There had been several top-secret meetings on the possibility since the end of 1997, and some delicate aspects had already been decided during discussions between top management teams. The representatives of both companies know each other well. The two firms were already collaborating successfully on fuel-cell technology.

Jürgen Schrempp was playing for very high stakes during these months. For what the head of Ford could not have known when he pushed to maintain the Ford family's majority holding in the merged company—a condition unacceptable to Schrempp—was that that the German was following a crafty and duplicitous strategy, pursuing Chrysler at the same time as negotiations proceeded with Ford.

And Trotman's mistake thereby was as simple as it was serious: He had underestimated Schrempp's confidence—or, perhaps, his megalomania!

The fact was that Daimler-Benz was also a definite second-best for

Robert J. Eaton. The head of Chrysler had never made any secret of his admiration for the Bayerischen Motorenwerke (BMW), but the BMW fortress remained impregnable even for the mighty American.

Making up for this, Daimler's chairman made the irresistible offer of having two chief executives at the new corporate giant. That was the final enticement for a marriage between the mass producer and the luxury automobile maker, which promised the conquering of world markets, power without limit, and incomparable financial gain.

The final merger talks were held in April 1998. One month later Schrempp announced the amalgamation with Chrysler to a surprised public. And Ford's Alex Trotman lost not just his "preferred candidate," but also the battle to become the number one premium automobile manufacturer.

Schrempp flew straight to London after the supervisory board meeting at which the announcement was made. The next day he challenged the competition in front of the assembled world press, saying provocatively that "we could take anyone on" and that "we would be the most innovative competitor, we would set the pace for the industry." He confidently promised the conquest of new markets.

With the announcement of the Chrysler-Daimler merger, it was clear that both "Project Alpha," a plan involving cooperation with General Motors, and "Project Beta," the proposed merger with Ford, had come to nothing. And now both Trotman and General Motors's chairman Jack Smith would really have to be on their toes, for Jürgen Schrempp would be content only when he was the head of the most profitable and most powerful automobile company in the world.

Those affected by the merger wanted to know where this journey was heading and whether they would all be on board when it ended. Would factories be shut? Jobs eliminated? Schrempp saw nothing but advantages, and he reassured people that the merger with the American car company was "not a restructuring merger" involving the closure of factories and the loss of jobs. In fact, the DaimlerChrysler boss averred, "the opposite is the case." Chrysler and Daimler were ultimately "highly complementary," which was also demonstrated by Chrysler's European sales of only a hundred thousand vehicles. "This number can, of course, be increased considerably with the help of our infrastructure," projected Schrempp; the converse also applied "for us in the United States." In effect, "we will create jobs, not destroy them."

Cochairman Eaton sent the same message, that new jobs would be created on both sides of the Atlantic. What he perhaps should have said is that more would be created than lost, for the oft-promised "synergistic effect" would at the very least lead to savings in the development departments and in administration, and perhaps at Dasa, too.

It was also necessary to look at the negative side. The rampant "mergeritis" of the nineties evident among banks and the chemical giants had increased profits but reduced the number of employees. Almost half of all company amalgamations in the United States had been failures. Elephant weddings most often led to confusion. Ken Hodge, manager of Mercer, a consulting firm, believed that Daimler and Chrysler would primarily have to face the unresolved problems of integrating company cultures, apart from having to deal with communication difficulties.

However, it was possible that the chances of the DaimlerChrysler deal were significantly better than those of other amalgamations, for the manner in which the managers of both automotive companies prepared the merger and led the company since then left room for optimism. Schrempp, certainly, saw it as "a marriage made in heaven." However, this marriage was nowhere near as straightforward as it seemed. And if Eaton had not softened his stance at the decisive moment of choosing the new company's name, this marriage made in heaven would have become enmity on earth.

The Power of One

We both thought that it would be sensible to have two at the top
during the initial period.
Jürgen E. Schrempp, Chairman of Daimler-Benz AG

This is an amalgamation of equals.
Robert J. Eaton, Chairman of Chrysler Corporation

This is a merger of two strong, healthy partners
that therefore have equal rights.
Jürgen E. Schrempp, Co-chairman of DaimlerChrysler

While the Schrempps were using their barbecue on their South African farm, their son Sander occasionally tossed bones over the electrified wires that surround the residential area. This upset the farm administrator, Jo Puck, who was aware that the smell of grilled meat would attract the lions and leopards that roam the area hungrily at night.

But bones do not fill them up, so the predators long ago developed the technique of hounding their prey into the electrified park fence, much to the displeasure of some.

Driving his Mercedes 290GD around the private grounds east of the residences, at all possible (and impossible) times of day, is one of Schrempp's greatest pleasures. The trail snakes through the terrain, and the many potholes allow the vehicle to progress only slowly. But time is of no consequence there.

One particular evening Daimler's chairman again threw all caution to the wind, instructing the driver to return to the farm without him. He set off alone for home, armed only with an Austrian Glock nine-millimeter pistol. It was August, the peak of the dry season, and there was hardly a green leaf to be seen. Elephants, in particular, cause havoc in this season of hunger, scratching the bark off trees with their tusks or ripping up decade-old trees to get at the relatively soft and tasty roots.

Anyone daring to wander across the land without the protection of a jeep or well-armed bodyguards must either have a devil-may-care approach to life or be tired of it. A two-and-a-half-mile trek through the bush seemed an endlessly long way compared to walking along an asphalt road.

At home, discontent and a certain uneasiness spread when the jeep returned to the farm without Schrempp in it. After the driver explained what had happened, a worried Sander climbed into his vehicle and set off to find his father. He too is one of those who enjoy being out at night. He sometimes tours the world of the savanna until four in the morning—but in a car and equipped with night-sight equipment.

Renate Schrempp was mad at her husband when he finally showed up. Consideration, even toward himself, is not exactly at the top of his list of priorities. Schrempp loves nature and he loves the risk. "Everybody knew that one doesn't go for walks there," declares Schrempp's business friend Theo Swart. But Schrempp went out there, without any bodyguards, "to recharge his batteries."

Those batteries were fully charged when the audacious German—having meanwhile advanced to become one of the most powerful of Europe's automobile executives—took the greatest risk of his career many years later.

Schrempp phoned the chairman of Chrysler in January 1998 with an extremely interesting offer. Robert Eaton was a little short with Schrempp at first, since it had been just two and a half years since exploratory talks took place between the two companies—without success. But this time there was more at stake. This time it was top secret. And this time only four top managers would be let in on the deal initially, not a whole squad like last time.

A merger of the two companies? Eaton was very interested when he heard what it was all about. Even if he did not agree immediately and asked for some time to think about it, Chrysler's chairman thought the offer was "great," as he freely admits now.

Schrempp flew to Michigan a few days later. The dialogue that took place while the Detroit Motor Show was going on—the first direct meeting to discuss merger plans—lasted just a quarter of an hour. Twelve further meetings would follow.

The messages were in code—"Denver" to "Cleveland," Daimler to Chrysler—as the top-secret "Project Gamma" took shape. The first meeting on neutral territory, in Geneva, quickly followed others in London and New York. At first only Gary Valade and Eckhard Cordes were involved in addition to Schrempp and Eaton.

The negotiating partners changed locations and tried to maintain anonymity. The cloud of smoke from Cuban Cohiba cigarillos betrayed a flurry of activity in grand hotels and luxury guest houses. It was not all peace and harmony when these two strong characters met.

In the end a friendship sprang up between the two leaders and is still going strong. When Jürgen Schrempp says to Robert Eaton in one of their joint presentations, "It's your turn, Bob," there is warmth in his words. He thanks his "colleague and friend" for his proven closeness. The trusting relationship was vital, for the two top managers were to lead the giant concern in close cooperation. Each of them would get a second desk: Jürgen in Auburn Hills and Bob in Möhringen. And solicitous as Schrempp was, he had "already furnished an office in Stuttgart" for the American.

Right from the start experts had no illusions about what type of amalgamation this would be. The analysis by *The New York Times* represents the views of many: "Daimler-Benz intends to buy Chrysler." On paper a merger of equal partners, in reality the deal was seen as Europe's largest industrial company swallowing a member of the Big Three.

However, Chrysler was by no means just the little brother from Michigan. While Daimler sold a good 900,000 automobiles worldwide in 1997, Chrysler sold 2.9 million. And while the Germans could report a net profit of $1.7 billion in the same year, the U.S. concern made $2.6 billion. In their own way both companies were extremely profitable.

Schrempp got the last laugh, however. He arranged that the decision about who should take charge was to be made by comparing the share balance. Every Daimler share was to be exchanged on a one-for-one basis for stock in DaimlerChrysler. Chrysler's shareholders clearly lost out: As the result of a complicated formula worked out as part of the deal, their shares were worth exactly 0.547 of a new DaimlerChrysler share. Thus existing Daimler shareholders would hold the majority of the new automobile giant. Schrempp had organized everything just right—from his point of view and in the interests of the Stuttgart side.

Schrempp's rhetorical knots have to be dissected with care to discover his real position on one of the key issues of "Project Gamma." *Der Spiegel* asked Schrempp whether he would consider agreeing to a merger with Chrysler even if Daimler-Benz AG ended up with a minority holding. "Never, no way, over my dead body"—that would have been the true answer.

Instead, he replied that Daimler's negotiating team had "taken all the factors into account" for this merger: market capitalization, the price/earnings ratio, and the earnings profile of each of the companies. "We agreed that on the basis of these evaluation criteria . . . Daimler's shareholders would hold 57 percent of the new company and Chrysler's shareholders 43 percent." It sounded really strange when Schrempp followed this up by saying that this was a merger between "strong, healthy partners" that have "equal rights."

On May 14, 1998, calls of "At last!" could be heard all around the boardroom as Hilmar Kopper arrived half an hour late. The chairman of the supervisory board excused himself, saying that he had been stuck in a traffic jam. But despite the delay, there was a general feeling of boisterousness, and not just on the part of the shareholders' representatives. "We were a hardworking supervisory board," one of the members comments, chest jutting out proudly, and not totally without reason. Not all the members could be present for this meeting, eight days after the first extraordinary meeting, because it had been scheduled at such short notice. But at least the absentee ballots of Manfred Schneider, Roland Schelling, Karl Feuerstein, and Manfred Göbels had been submitted for this historic decision.

Schrempp was being mischievous when he remarked to the assembled members that there was really nothing more to discuss. Or was there? If any board members wanted more information, all they had to do was ask. But those present already knew everything, said Schrempp, shrugging and appearing not at all willing to present the usual situation report complete with overheads, factual presentation, and forward planning. Schrempp was so euphoric that any criticism seemed irrelevant. Daimler and Chrysler would merge, share values would climb sharply— the background had already been explained the week before. Everything else was a mere formality, wasn't it?

A certain skepticism, however, was noticeable during the meeting. This was hardly surprising. For after the lofty thoughts of the first few days, questions were at last being raised that even Daimler's chairman did not find so easy to answer.

There was the battle of the cultures. "Is it actually possible to mesh the present diverse corporate cultures at all?" one of the bank representatives wondered, and the question was justified. Could anyone imagine the Americans eating schnitzel or the Germans being happy with American fast food? Whatever the case, in the future they would have to sit at one table, for after the two shareholders' general meetings and the merger that would be decided on there, U.S. board members would also participate in Daimler-Benz supervisory board meetings at first.

Schrempp repeatedly laid out his belief that all was well: "There will be no rationalization effects." And, using all his powers of persuasion, he issued reassurances that "no rationalization measures will be undertaken, either."

The chairman of the Debis central works council, Herbert Schiller, had other concerns: "What will happen to this relatively small service provider?" The concert performed by the two automotive giants threatened to drown out little Debis.

"And what would be the future role of aviation and aerospace?" Peter Schönfelder was clearly worried. The chairman of the central works council at Dasa's Augsburg factory was seconded by Hubert Curien, France's former minister of research and technology, who, like many of his countrymen, was less than enthusiastic about U.S. capital. The consequences for the European aviation industry's harmonization process would be considerable were the French government to oppose privatization of Dasa's partner, Aerospatiale. Supervisory board members feared that parts of the military sector, or all of it, could even be up for grabs soon. The company film promoting the merger, which was being shown all around the world, mentioned only Airbus and Eurocopter. Was this deliberate? Manfred Bischoff maintained his silence throughout the meeting.

Could the aviation and aerospace division survive when the automobile sector was so heavily reinforced? A good tenth of the workforce of the new DaimlerChrysler AG was still active in Dasa and Debis. In view of the impending threat caused by a reduction of the divisions' significance, some employees, particularly at Daimler-Benz Aerospace, were extremely worried about their future. In Great Britain the value of British Aerospace shares had risen sensationally. The *Financial Times* speculated that it would now be easier for Daimler "to relinquish control of its aviation and defense interests." A rumor about a Dasa sale was even circulating around the supervisory board.

Schrempp would hear nothing of all this: He still wholeheartedly supported the future of the aviation and aerospace sector. And when one of the members wondered out loud about the future of Manfred Bischoff, Kopper countered skillfully, saying that Bischoff would still have responsibilities in the future—though whether these would include being head of Dasa was not clarified.

The chairman of the board disposed of all doubts in typical Schrempp fashion, saying that there was nothing to be concerned about, that all these questions had already been cleared up with the U.S. partners. Chrysler's management, Schrempp assured those around the table, had accepted the continuance of both the service-providing business and the aviation and aerospace division—at least as long as they remained profitable. For the moment no pressure could be expected from the U.S. management side.

The partner from across the Atlantic had even promised that Daimler could use its contacts with Lockheed when launching the new Superjumbo A3XX onto the market. In the end, a rather amused Peter Schönfelder teased that "Jürgen E. Columbus" simply could not wait to set off to conquer the New World, come what may.

Finally, the unanimity that Jürgen Schrempp had desired prevailed, and the supervisory board added their vote to that of the board of management in favor of the merger between Daimler-Benz AG and the Chrysler Corporation. Schrempp's transcontinental transport concern began to take shape. The campaign of conquest initiated by Germany's Columbus moved inexorably forward.

The only employee representative to be informed in advance about the course of the negotiations was the seriously ill Karl Feuerstein. Schrempp was able to obtain his vote relatively easily as DaimlerChrysler AG was a German company with its head office based in Germany—a bitter pill for the U.S. management team to swallow, but for IG Metall a precondition for successfully exerting influence. So ensuring the employees' vote in favor of the merger was merely a formality.

After all the hopes and promises, there was still one significant hurdle that remained to be overcome by Daimler's supervisory board. As late as the summer of 1998 it was still not clear whether the merger would really take place or not, even if the German monopoly commission and the U.S. Justice Department's antitrust lawyers agreed to the move, for even the most attractive business needs rethinking if fundamental preconditions are not fulfilled.

What, then, would be the new supervisory board's lingua franca—German, English, or perhaps one of the German dialects? Communication was certain to be a problem, for on the employees' side alone, at least three members of the supervisory board could hardly speak any English.

Robert Eaton had already proved that he was quite capable of heading a company in tandem with an equal partner. When he surprised everyone in March 1992 by switching from General Motors to Chrysler, he was initially given the position of vice chairman. After pressure from the board of directors forced Chrysler boss Lee Iacocca to hand over his post to Eaton the following year, his passage to the top was ensured. Eaton prevailed over Chrysler president Bob Lutz in much the same way that the duel between Schrempp and Werner was resolved, despite the eventual loser seeming to have held far better cards at the start. Lutz, a Swiss, was at his internal rival's side for four years, and the management team of "Bob and Bob" developed Chrysler into a technologically innovative and financially lucrative company that was considered to be one of the pearls of the U.S. automotive industry.

Nevertheless, it was the German who finally gained the upper hand—as Schrempp and Eaton's joint appearance onstage in London clearly demonstrated.

The London Arena is a modest building in the heart of the Docklands, designed specifically for rock concerts and sporting events. On May 7, 1998, a fleet of luxury Mercedes cars accompanied by a single Chrysler limousine stood outside the arena. Inside the lighting was subdued—only the new company logo was illuminated by spotlights.

The heroic entrance of the gladiators followed. They were enthroned on a raised dais above all the rest. A dozen TV cameras, an army of photographers, and more than two hundred journalists from all over the world documented the scene for the portrait gallery of neoliberalism.

First Schrempp (who else?) described the advantages of this megadeal. Then Schrempp (who else?) discussed the global battlefield of the intensified automotive war. Finally Schrempp (who else?) promised profits of unheard-of proportions. Then, at last, his friend Eaton could say a few words. No, Chrysler was not a junior partner; like attracts like, he said. He had found his "preferred partner"—he meant Daimler-Benz, not Jürgen Schrempp, who had, it seemed, set him straight about who was to head the DaimlerChrysler household, first behind the scenes and then before the assembled world press. And initially Eaton accepted his fate with an astonishing degree of willingness. In three years, according to the American, "Jürgen Schrempp will be the sole chairman."

The journalists would have been forgiven if they had asked, "Why wait so long?" Right from the start Schrempp showed his friend Bob the lay of the land. Schrempp had never in his whole life tolerated having someone share his power—at least not for any length of time, and certainly not for as long as three years.

In the same week, resistance to the takeover of this established U.S. firm by the German automobile and aviation company increased in the United States. For one thing, angry shareholders argued, Chrysler was being sold much too cheaply. Furthermore—and this was where Schrempp's victory and Eaton's defeat weighed most heavily—the new giant company was to be registered in Germany. It would take Eaton a long time to live that one down. The foreign registration was also the reason why the new DaimlerChrysler shares, with the stock market symbol DCX, suffered their first industrial accident even before the start of trading in DaimlerChrysler stock on November 17, 1998: They were left out of the S&P 500 index. Standard & Poor's based their refusal on the fact that the company was now located outside the United States. Schrempp characterized this as "incomprehensible," but he tried to play down the setback, commenting that the decision was "shortsighted," as "this important share index" now no longer included the third-largest automobile producer.

Formally, the model for the new company envisaged a double command into the new millennium. Schrempp and Eaton were to manage the company's business together until 2001, and then the American was to move to the supervisory board. Before this there would be no division of duties, as both were responsible for everything and each would be free to meddle in the affairs of the other. It certainly seemed as if this plan meant that conflict was structurally preprogrammed into the new concern.

David E. Davis, chief editor of *Automobile* magazine, was pessimistic: "The Second World War would be refought" on a daily basis, "sometimes in one department, sometimes in another."

One man's power was another man's powerlessness. What naive desire, what infantile simple-mindedness, what trusting faith in God made Robert Eaton even begin to hope that he might have the glimmer of a chance of having an equal say in the running of DaimlerChrysler AG? Of course Schrempp would have him beside him on the stage. Of course the photographers would get a picture of them together. Of course Schrempp would keep himself on a tight leash for the first year. And of course Bob could take his allocated place beside his friend Jürgen.

But did Eaton really believe that he would be able to manage Daimler-Chrysler on an equal footing with Schrempp? Did the top Chrysler manager really assume that Schrempp would tolerate managing side by side with anyone? Such ideas could only be based on ignorance of Schrempp's career, which speaks the language of the manager's bible: Thou shalt have no other business gods but me.

The German was well aware that since the DaimlerChrysler deal he was now more than ever in the spotlight of the world's press and a critical American public, who were carefully watching every move he made. If only for this reason, he had to be much more careful and proceed with far more tact.

This was difficult for Schrempp, but he would succeed in keeping himself under control. If he wanted to achieve his aim at the beginning of the millennium—and there was no doubt that he did—then he would just have to be patient this once. Until one person alone held all the power in his hands.

Not two months after the shareholders' overwhelming agreement at the separate general meetings on September 18, 1998, Schrempp received the first accolade for the megadeal: A jury of eight distinguished experts and the chief editor of the German business magazine *Manager* voted the "Stuttgart powerhouse" "Manager of the Year," way ahead of the rest of the field.

Jury members justified awarding him this, the most prestigious business prize in Schrempp's homeland, on the grounds of his "pathbreaking decisions," his "unwavering assertiveness," and the fact that he "led the company back to its core capabilities and knocked it into shape for the Chrysler merger." Schrempp was the "German CEO for the global dimension."

An unbroken series of national and international prizes was to follow. The German's success was also thoroughly acknowledged in the English-speaking world. *Automotive News* crowned him "Industry Leader of the Year." The magazine *Business Week* included Schrempp (though not his cochairman, Eaton) in its "Top 25 Executives of the Year." The decisive factor in this honor (which was granted only to two Germans—Schrempp and Volkswagen chairman Ferdinand Piëch) was, apart from the successfully implemented DaimlerChrysler merger, principally for the fact that Schrempp had "quieted the naysayers"—at least for the time being, added *Business Week* farsightedly.

A good decade and half had passed since Schrempp, aka "Mr. Clueless," set off in October 1982 to conquer the New World. But now Schrempp was at the peak of his career. With the takeover of Chrysler and his preprogrammed sole control after 2001, media reports were falling over themselves elevating Schrempp to a "superman" or "superstar."

The former Mercedes apprentice from southern Germany today controls the most powerful industrial company on the European continent, the world's number three automotive concern. He has long been one of the driving forces of the globalization process. *Quo vadis,* Jürgen E. Columbus?

Under the sign of the Chrysler star, Schrempp has begun forging the first, and up to now only, German company with worldwide influence. But it is an explosive mixture that is united in Schrempp: courage and recklessness, genius and megalomania lie in close proximity within him, as close as the dangerous ease with which the use of power could become abuse of power.

The German manager is playing monopoly on a global scale. "This amalgamation will rewrite the rules for the automotive industry," Schrempp has insisted. With this megamerger he has indeed rewritten the rules, and at the moment he is winning. But what if Schrempp loses his global game? Then more than just a few hundred thousand jobs at Daimler and Chrysler are at stake.

Operation Moneymaking

Thus he earns 600 marks an hour—regardless of whether he is in
bed or at a Board meeting.
Alexander Dauensteiner, on Schrempp

Schrempp has always hated the topic of his salary, openly raised every
year at the shareholders' annual meetings by the Dachverband der
Kritischen Aktionäre (the umbrella organization of critical shareholders). Year after year the answer has to be found in the publications of the
business press, often only as an estimate. For the DaimlerChrysler chairman has never yet been prepared to provide his shareholders with information on his income from the company.

It certainly is an uneven match that is played out day by day. One of
them, with a salary of about $3 million, dictates the play. The other, with
almost $11.5 million a year—which doesn't even land him on the top
floor of better earners in the United States—nods encouragingly at his
doubles partner.

Schrempp, usually so aggressive, learned from the sick-pay issue that
salary can be an explosive issue. So this time he left it to his friend Bob,
who was usually allowed to play second fiddle at best, to broach the subject of a pay raise for his German cochairman with this statement: "I
could well imagine that in the future Jürgen would earn the same as I do."
Thus Schrempp's salary must be "brought into line" with Eaton's own.

If Europe's best-paid chairman of the board achieves such an exor-
bitant (by German standards) increase in his salary, it will send a signal—and not only in Germany. For then it will be just a question of time
before the chairman's salary at other large European concerns would go
up on a similar scale.

The Americanization of Schrempp's salary became the topic of a heated nationwide debate ahead of the shareholders' general meeting of 1999.
Hence the board could not prevent the matter from playing a central role
at the very first DaimlerChrysler shareholders' general meeting, despite
the fact that the question of raises in salary was not even on the agenda.

At this meeting, on May 18, 1999, Robert Eaton—who previously could greet personally almost all of the 150 (at most) participants at a Chrysler shareholders' meeting—spoke before a world-record audience of almost 19,500 shareholders (about 500 of them from the States) in Stuttgart's Hanns-Martin-Schleyer-Halle. His English was simultaneously translated. The automobile man at the lectern appeared surprisingly unconcerned as he announced the "report on the state of business of the DaimlerChrysler Group." Eaton's tone was strangely dry as he offered up the new figures: sales records in the vehicle sector with an increase of 12 percent; and a record operating profit, up by a fantastic 38 percent. Then, after thanking Eaton, Schrempp took the floor.

At this shareholders' general meeting, the allocation of roles (and weight) within the concern became obvious not just to the American side: Eaton was to evaluate the present, while Schrempp was to decide the future. Eaton's clock would soon run down, and then Schrempp's hour would come. And while Eaton would soon be sent off to the supervisory board, Schrempp's motto was to implement "speed, speed, speed" at all levels and at all the factories of the new worldwide company. The autocrat-to-be was already dreaming of "the car beyond the immediate future," largely "defined by networked electronics." And until the year 2002 DaimlerChrysler would bring one new passenger or commercial vehicle model onto the market every month.

Schrempp played down many of the problems, such as the European Commission's complaint that Daimler was hindering competition: It was a matter of "just ten individual cases." Others' concerns were not even seriously discussed.

The corporation had so far invested a good $1 billion in constructing its new Smart model, taken over the 19 percent participation of the Swiss Nicolas Hayek, and founded its own development company at the end of 1998. But the figures were terrible. And at the analysts' conference at the beginning of May, Schrempp was still threatening to undertake "drastic measures" if the Smart concept did not get off the ground soon.

A few days later, however, in front of the shareholders, the chairman of the board whitewashed the disaster with the city car with a reference to the "clear trend of the last few weeks." The Smart—which, at the end of 1999, was "on its way out," because of its poor image and even worse sales figures, according to insiders—was worth a second chance.

The negative development of the transcontinental ownership structure apparently gave Schrempp no grounds for concern, either. (The proportion of U.S. shareholders had sunk from about 44 percent to 25 percent since the merger, the result of keeping the DCX apart from S&P's 500 Index. This was catastrophic for DaimlerChrysler shareholders in the U.S.) And instead of critically examining the migration of leading personnel to competitors such as General Motors or Ford, Jürgen Schrempp resorted to phrases along the lines of "We will be satisfied only when the cream of the crop are standing in line for the chance to work for DaimlerChrysler."

The division of duties at the shareholders' general meeting (May 1999) was arranged in advance: Schrempp gave his critics empty replies, Eaton made exemplary comments on the questions passed on to him by Schrempp, and this time Hilmar Kopper had responsibility for dealing with the question of salaries. The supervisory board's chairman had already done a good job ensuring that the issue of the Americanization of the chairman's salary would not be picked up as a central theme for discussion by the supervisory board itself, but rather farmed out to the general committee. Even if this procedure was undoubtedly in violation of the rules, there was a strategy behind it.

The employees' representative was the one who raised objections, within the union framework, to the intended raise for the top manager of DaimlerChrysler AG. "We were still against it until recently," says one supervisory board source. But the topic had been banished to the general committee, where a strong union representative had been missing since Karl Feuerstein's resignation on health grounds. And his successor, Erich Klemm, the new central works council chairman and deputy chairman of the supervisory board, had been in office for only a few weeks.

"We now have to rely on Erich," said one of his colleagues on the supervisory board, still hopeful at the time.

Klemm, head of the Labor council for ten years at Mercedes-Benz' largest plant, in Sindelfingen, has long experience with industrial labor relations. He is less concerned about the excessive increase in the remuneration of German board members than about the level of employees' incomes. "Of course we could see the connection between a possible increase in the chairman's earnings," says the forty-four-year-old employee's representative, "and what, for example, the workforce receives in profit sharing." (Profit sharing, while common in the United States, is still rare in Germany.)

The new man at the head of the metalworkers in the Daimler-Chrysler concern has no wish to block the explosion of management salaries. Instead he believes that the employees must "get more out of it" than adjusting the profit sharing to the operating profit of the old Daimler-Benz AG.

Furthermore, he does not believe "that there will be any increased Americanization as a result of the merger." But who can serve as a check on the chairmen of large concerns if not the supervisory boards that have been put together expressly for this purpose? Kopper & Co. could at least have expected a lot more resistance than they seemed to be getting.

The state of DaimlerChrysler's supervisory board in the period following the merger provides much food for thought. Whereas meetings before the merger began at ten in the morning and mostly lasted around seven hours, nowadays lunch is eaten together first and a shorter program, often lasting just two hours, is reeled off. The difficulty of having the U.S. participants attend the meetings is cited as one reason for the change.

Whatever the case, Schrempp, Eaton, and Kopper can be happy about the end result: They can bustle around to their hearts' content. The comment by Peter Schönfelder, IG Metall's representative on the supervisory board, is short and sweet: "A full stomach doesn't put up much resistance."

As if this were not enough, when asked about how communication between the Germans and their American colleagues was proceeding, one of the former replies: "I know all the faces after a year. But I could not put a name to all of them." And this one year after the merger!

"How much do you earn at the moment? And how much will it be in the future, Mr. Schrempp?" The questions posed by the shareholders' spokesman, Alexander Dauensteiner, regarding the first point on the agenda, the vote of formal approval of the chairman, at the shareholders' general meeting of May 1999 came as no surprise.

Yet again Jürgen Schrempp proved to be a clever tactician. Instead of directly confronting the shareholders with figures showing that the board members' salaries had quadrupled or quintupled, Hilmar Kopper, in charge of the meeting, presented a four-component strategy. "The new system of remuneration," explained the Frankfurt banker, "is the same for all members of the board—regardless of their nationality and work location." Which did not mean that every board member would receive the same salary.

A flexible system was introduced for 1999 by means of a relatively low fixed remuneration, varying bonuses for each of the members of the board, a three-year performance plan, and share value enhancement rights. A new stock options plan will follow in the year 2000, whereby Schrempp and his men will profit directly from any increase in the share value. Hence, year after year, the top managers are guaranteed compensation in the tens of millions.

It is pointless to mention that the questions of pay raised by the critical shareholders group remain unanswered this year, too.

The earnings battle has basically been won, its implementation only a question of time. And if savings are not to be made at the very top, then they must at least be spread across the workforce.

When IG Metall, after many years of stagnation and real losses in earnings, first demanded a wage raise in the fall of 1998, Daimler-Chrysler's chairman bluntly refused: "The IG Metall demand for a 6.5 percent wage increase is sending the wrong signal." And even when the compromise offered by the union's negotiating team was fixed at 3.2 percent, the chairman of the board expressed dissatisfaction. Schrempp really does know when it is necessary to keep a tight grip on the Group's finances—and when it isn't.

Why, however, did no one come up with the idea, banal and effective in equal measure, of linking the chairman's salary to the average employee wage? An increase of a little over 3 percent, which is what the workers were asking, would allow Schrempp to afford a new fourth car—if not a fourth villa.

A multiplication of salary on the American principle with a simultaneous refusal to provide real information in accordance with German secrecy? No manager before Jürgen Schrempp has dared make such a self-serving interpretation of the legal framework. The Daimler boss skillfully picks the raisins out of the transatlantic cake.

Despite this, only a minority of the participants at the shareholders' general meeting protested. This was probably due to the fact that the German shareholders were happy with the 30 percent rise in dividends.

One of the voices raised in dissent at the stockholders' meeting, though not about the salary issue, was that of Evelyn Y. Davis. Active at the general meetings of about forty U.S. companies every year, she complained angrily about the obvious takeover of the Chrysler Corporation by Daimler-Benz and about the unequal division of power at the top of the new company. "Having two captains on the bridge is like having two chefs in the kitchen," she commented, and suggested that Eaton would do better to move to General Motors. Her criticism of the German CEO was just as clearly expressed: "Mr. Schrempp, why did you buy Chrysler? Wouldn't a joint venture have been enough?" To this day, the vigorous Davis is still waiting for an answer, so she supplies one herself: "The Germans want to take over our high-tech industries." Though he may have remained silent on this one, Schrempp will certainly have to become accustomed to vehement objections like these.

In regard to salaries, Schrempp heard another loud critical voice, too, this one from a different corner—one that Schrempp himself would have considered a most unlikely source.

The Schrempp familial well has been poisoned since at least February 1999. In her letter to the daily newspaper *Badische Zeitung*, published in Schrempp's hometown of Freiburg but quoted throughout the nation, the former wife of his elder brother Günter gave free rein to her anger. DaimlerChrysler's top floor has been considering whether to have part of the board members' wages taxed in the United States in the future, where, by making use of the advantages offered by the U.S. tax laws (with a top rate of only 43 percent), "Operation Money-making" could be further refined—at least this is Gerlinde Schrempp's allegation.

But the planned tax transaction may have been taking things a step too far. Gerlinde Schrempp's letter was addressed to her ex-brother-in-law personally: "Dear Jürgen," wrote the angry single mother, "I simply could not believe that you, and your colleagues on the DaimlerChrysler board who were also not badly paid, have forgotten where your training was made possible." Which country arranged "for a successful manager's career such as your own"? "That is surely not the same United States that could look forward to receiving your tax payments in future?"

Furthermore, continued his former sister-in-law, "people like you do not have any additional expenses" when they go abroad. "I believe that not a cent has to be paid out of a small wage there, or am I wrong?"

Her criticism became a bitter accusation: "Terms such as 'greed,' 'could never get enough,' and 'forget one's roots' occur to me." In her disappointment she reminded Jürgen of his "dear, honest father," who had been "particularly proud of you." If, however, he "were alive today he would be ashamed of your plans."

Such discussions about salaries and their being taxed abroad do not affect Chrysler's managers. They have long since made arrangements to ensure that they will not have to rely on a welfare check in the future.

Arranged in 1995, with the creation of executive officers in the company, "continuation agreements" guarantee that in the case of dismissal "for an important reason," a settlement amounting to "two to three times the basic [annual] salary" and further bonus payments are forthcoming.

And with the amalgamation of the two automobile companies, Robert Eaton has also been promised "an improvement in pension provision." In the best of cases this would, as established in the joint report on the merger compiled by the members of the board, come to an impressive $30,000 a month.

If Chrysler's management is given notice within two years of the merger, they will receive lucrative settlements, here estimated as a lump sum: Gary Valade, $4,601,383; Thomas Stallkamp, $5,487,445; and Robert Eaton, $24,435,997. And as it can be assumed that Jürgen Schrempp is keeping an eye on the company's finances (as well as his own, of course), the premature departure of Eaton before the summer of the year 2000 can be ruled out.

Still, these sums are peanuts compared with what Chrysler's top management made from the companies' merger, in addition to the millions they earn in salaries. Calculations can be based on the information about Chrysler share options obtained on May 27, 1998, according to which Thomas Stallkamp received 379,384 DaimlerChrysler shares, James P. Holden 407,771 shares, Gary Valade 442,685 shares, Robert Lutz 683,380 shares, and Robert Eaton 2,267,579 shares. Taken together, the executive officers, who number about thirty in all, received 8,521,319 shares within the framework of the merger. A further 3,533,234 Daimler-Chrysler share certificates must be added to this sum for options that were taken early. And the value of the stock options, of which the top five managers possess about one-fifth, rose by more than a billion dollars as a result of the merger.

A lot of questions remain unanswered when it comes to analyzing the impact of the merger, but when it comes to top executives, especially the Chrysler managers, one thing can be definitively said: Nobody is worse off than they were before.

CHAPTER 9

Winds of Change

Daimler-Schrempp: We've Done It! Best Year Ever!
Headline in Bild

At the April 1988 press conference announcing the company's annual report for the previous year, Schrempp fiddled nervously with his pen for two hours straight. Bedeviled by a constant inner restlessness, the man vented his excess energy on the writing implement. At the next question from the assembled journalists and TV reporters, he again pushed his spectacles back into place and made a short-lived motion to stand up, only to immediately suppress this impulse once more. His actions were reminiscent more of a child who is incapable of sitting still for two minutes on end than of a fifty-four-year-old man in a leading position.

Given the figures he had been able to present in the second year after taking over from Edzard Reuter—total sales rising from $56 billion to $66 billion, a record dividend of 85 cents, and a surprise one-time extra distribution of $10.65 per share—Jürgen Schrempp had every reason to be unrestrainedly jubilant, and one would not have been surprised to see him beside himself with joy. But the final spark of pleasure was missing. Not because Schrempp was dissatisfied, but because he expected more of himself and his employees—far more. Hence his prediction that "next year we will achieve somewhere around $71 billion, and the year after next $85 billion." This was a "conservative" estimate: "We calculate that total sales will double to $133 billion within ten years."

When the press conference presenting the annual report came to a close and the scribbling guild had departed to work in the press room or have lunch, the Lord of the Stars was still patiently answering and lecturing a large number of TV reporters. A few questions, however, were not answered and still remain unresolved.

"This concept allowed us to create more than 12,000 jobs in 1997" was one of ten key quotations handed out to media representatives in a position statement by the chairman of the board before this particular press conference. The next day there were regular references to "new" and "more jobs" or the "new employment trend" scattered among the laudatory phrases and in headlines. There was nothing to quibble about in the statements themselves: In 1997 Daimler-Benz created 10,039 jobs while elsewhere in Germany jobs were lost. This was more jobs than at any other German industrial company, since in reality an additional 2,000 should have been added to this sum as a result of consolidation measures. With success in the marketplace and improved competitiveness, "more new jobs" would come on top of this, announced a thoroughly satisfied board member. This positive employment trend has continued with the merger between Daimler-Benz AG and the Chrysler Corporation. In 1998 alone, the year of the merger, Jürgen Schrempp and Robert Eaton were in a position to announce expansion by almost 16,000 jobs, to more than 441,000 in the entire Group.

So much for the ideal Daimler world with its lovely references to the preceding year. The Lord of the Stars was less willing to discuss the extent of job losses in previous years. Employment at the former Daimler-Benz AG reached a peak in 1991, with almost 380,000 employees. In the following years the layoffs carried out by the then chairman of Dasa, Jürgen Schrempp, led to a massive reduction in the workforce.

When Edzard Reuter left and Schrempp was enthroned, almost 70,000 employees had already been "released." When parts of the company were sold, roughly half of the workers found a new employer—the rest were given pink slips. And Daimler chairman Schrempp seamlessly continued where Dasa chairman Schrempp had left off, laying off an additional 20,000 employees by the end of 1996.

In year one of Schrempp's chairmanship of the entire Group the jobs situation reached an absolute low point. There were exactly the same number of employees at Daimler-Benz AG as when Jürgen Schrempp was made chairman of MBSA. All in all, the Daimler board got rid of a grand total of 89,223 jobs between 1991 and 1996, thus shedding almost a quarter of the workforce. Schrempp's celebrations meant nothing less than that under his leadership the company had simply improved in the rankings—taking the previous ten years as a whole, the company's jobs rating had improved from last to next-to-last.

The chanting of Daimler's jobs song sounds all too macabre for those affected by the cuts, when one contrasts balance sheet profits with the number of "new jobs created." While total sales in 1998 increased by 12 percent, operating profit surged by 38 percent, and profit per share shot up by 30 percent, the number of employees increased by a mere 4 percent.

What appeared to sound so melodious from an economic point of view was in line with an approach involving getting the maximum amount of work possible out of the minimum number of employees, and sometimes even more than this. In effect, DaimlerChrysler AG has exactly the number of employees that it needs to fulfill the cochairmen's profit targets. If the workforce were further reduced, customers' wishes could not be met and the competition would be the beneficiary.

One thing that can be said for Schrempp is that, in contrast to many others, he rationalizes on all levels of the company—even the highest. In September 1998, when calls for peace were echoing through the corridors of the DaimlerChrysler headquarters in Germany and the United States, Schrempp was already thinking of the forthcoming board structures that should exist when the long-running merger process reached its conclusion: "In the longer term" one must face the question of whether there should not be "one particular person on the board who is responsible for their own particular area." What sounded so innocuous at first was Schrempp's way of hinting at a coming restructuring of the DaimlerChrysler board of management.

Nine months later, the CEO was more concrete. The figure of seventeen board members was "very high" but reflects "our current structure." One did not need to have second sight to know what these words meant. Schrempp had already spoken once before of too many chiefs and not enough Indians. At that time, during the radical restructuring of Daimler-Benz AG, he put his own house in order as consistently as no one else could—and also at the highest levels. This time, in May 1999, he made no secret of the timing of what he was aiming for: "Today the company has the structure that it needs for the next two or three years."

The battles will have been decided by the year 2003 at the latest (the last year of the current board members' terms), and most probably earlier: For passenger cars, Jürgen Hubbert or Thomas C. Gale; for commercial vehicles, Theodor R. Cunningham or Kurt Lauk; for procurement, Thomas W. Sidlik or Gary Valade; for sales, James P. Holden or Dieter Zetsche—and there is no limit to the further permutations in the game of musical board seats.

In the end there will be a notably smaller and more efficient management committee of about ten members—and some of the directors will have left the company with a comfortable settlement.

But even Schrempp himself should not imagine that his seat is enduringly stable. After the phase of his shared reign with Robert Eaton, the German will take over the rudder as sole chief executive officer. But the days of Schrempp's unlimited say over the fate of the Daimler-Chrysler empire are numbered.

Behind the scenes the contest over his successor has already begun. Thomas Stallkamp and Eckhard Cordes—together with Jürgen Hubbert, Manfred Gentz, Gary Valade, and Schrempp (all members of the eight-man chairman's integration council)—are among the hottest contenders in the final battle for the chairman's post.

There is no doubt that the fifty-two-year-old Stallkamp, president of the company and the third man behind the top duo, has the better cards when it comes to advancing to the position of Schrempp's deputy when Eaton departs. In the key area of passenger cars, the integration teams report to him first and only then to Jürgen Hubbert, the director actually responsible for passenger cars. "One person must oversee the integration of the automobile division," Eaton says, championing his protégé, in an unmistakable message from the Americans to the Germans, "and that is Stallkamp."

On the other hand, the highly professional manner in which Cordes (who is five years younger than Stallkamp) so perfectly put together the

Daimler-Chrysler merger speaks well for him. But who, in the next millennium, will still be interested in the splendid merger management of an Eckhard Cordes, especially considering that Stallkamp was also very closely involved in it?

A lot will depend on how the man responsible for the Group's development does his job in the coming years. By buying back ABB's 50 percent share in Adtrans (originally a joint venture between Daimler and ABB), DaimlerChrysler has taken another step toward becoming a globally active mobility concern. But the world's largest producer of railroad traffic technology, for which Cordes bears responsibility, continues to lose money—and there does not appear to be any change in this trend. If the fortunes of Adtrans fail to improve, he may suffer the same fate as so many before him.

Cordes has recognized the danger, and has implemented a restructuring concept at ABB, cutting about 1,400 of the 7,600 jobs. But this move will achieve little, if the works councils are to be believed.

The price that will have to be paid for Schrempp's being sole chairman, and thus having complete control, for a limited time may be high. For an American could well follow the German as the head of the Group.

The U.S. side will hardly allow the boss's post to be, in effect, bequeathed to Germany. However, even if Stallkamp is promoting the idea that it "no longer has anything to do with nationality" or the "national flag," the concept of an American chairman of the board on the eleventh floor is, at least for the moment, difficult for some in the German headquarters to grasp. These skeptics are either incapable of coming to terms with it or choose not to.

In a few years, however, the Daimler world may well look different. Whatever the case, Stallkamp is deliberately playing down questions on this topic. "Of course" there is a possibility of his becoming Schrempp's successor. But the next chairman "could, however, also come from Botswana or from Switzerland," as he said when interviewed by Karl-Heinz Büschermann from the newspaper *Süddeutschen Zeitung*, for "it could be anyone."

Not really, Mr. Stallkamp. For in the end it is the results of today's top managers that will be the deciding factor. DaimlerChrysler's president has in any case set the bar very high, saying that only if savings of $3 billion are achieved for the year 2001 will the company operate properly. "One and one must make more than two," according to Stallkamp. If "less is achieved, they must fire us"—a clear statement for one who intends to become the chosen Lord of the Stars via the passenger car sector.

The well-being of the Group's employees plays a minor role in the global game of markets and money. The advances made by Schrempp's system for optimizing human resources can be judged not least by the current agreement on measures to reduce absenteeism at the Sindelfingen production works.

Management did not shy away either from a tiered bonus system based on the amount of sick leave taken, or from instituting meetings between employee and supervisor after the employee returns from sick leave. The aim of "reducing days lost due to sickness by 4 percent for manual workers and 2 percent for salaried employees over the next few years" hardly speaks the language of a socially engaged capitalism: Work, work, build that car, getting sick won't get you far! This is Daimler-speak for an ideal way to increase company value.

The catalogue of measures ranges from a campaign to increase awareness of the negative effects of high absenteeism to "consistent action" against employees with "noticeable" amounts of time lost. And so that everyone is clear about what is going on, bulletins about current absenteeism statistics are hung on information boards next to the current production charts—without the perpetrators being named, at least not yet. Those who miss fewer than four days annually are rewarded with a bonus of $154. Anyone who is often ill gets nothing and, as a persistent offender, could end up without a job. As they put it so well in the introduction to the Sindelfingen union agreement, the agreement was intended "for the employees' welfare."

Jürgen Schrempp thinks that all this is really not so bad. After all, "workdays that are lost for reasons other than illness harm colleagues who have to take over the work."

The Sindelfingen contract, which cannot be annulled before December 31, 2000, has been blessed and signed by Erich Klemm, chairman of the Sindelfingen central works council and a member of the supervisory board. *Quo vadis*, IG Metall? One gradually gains the impression that the men at the top of the most powerful trade union in the world do not want to admit to the continuing decline of trade union influence—or that they have already capitulated.

———————

As much as Jürgen Schrempp likes to mock the billions lost during the reign of his predecessor Edzard Reuter (which, of course, ultimately made Schrempp's rise to chairman of the board possible), he also managed to use that loss to the benefit of the Group's finances. Since 1995 the Daimler-Benz Group, now Germany's most profitable industrial company again, has been paying comparatively little tax as a result of the tens of billions incurred in losses.

Despite brilliant profits, Schrempp was able to withhold a full $1 billion from the Ministry of Finance in 1997 alone. As, under German law, profits that were retained by the company were taxed at a higher rate than those paid out to shareholders, financial director Manfred Gentz without further ado released the reserves to shareholders. This perfectly legal and aboveboard move meant that the treasury lost additional billions in potential tax credits. "If that isn't straight shareholder value . . . ," railed the newspaper *Badische Zeitung* in Schrempp's hometown.

Finally, in September 1998, after three long years of withholding taxes, Daimler's CEO appeared in public to announce the company's rediscovery of its civic duty. "We are paying a very large amount of tax this year, and by this," Schrempp continued, "I do not just mean transfer taxes due in connection with the amalgamation."

Nevertheless, the discrepancy between the Group's continuing record profits and the (perfectly legal) refusal to pay correspondingly large taxes has done lasting damage to the company's image. Talks between the financial director of DaimlerChrysler AG and the government's financial experts in the spring of 1999 were understandably stormy.

In a strongly worded letter to the German chancellor, Manfred Gentz pointed out that a plan according to which a certain share of the Group's profits originating abroad would be taxed a second time in Germany might well result in investors on the U.S. capital market losing interest in Germany's largest company. "Only those who offer attractive profits receive capital investment," went the letter, and this double taxation would impose "unacceptable additional burdens."

A few days later the otherwise levelheaded Manfred Gentz was quoted as saying that unless some tax relief was granted, the Group's headquarters and top managers might all seek to move their domiciles out of Germany. Gentz's threat resulted in utter astonishment at the Ministry of Finance, for the Group is not exactly one of the country's most prominent taxpayers. According to ministry spokesman Torsten Albig, Daimler had not paid any tax at all since 1995. Gentz's and Albig's statements both made headlines.

Of course Daimler had paid taxes—more than $58 million in commercial earnings tax in 1998 alone, retorted the company's press officer, Christoph Walther, angrily. He did not mince words: "The Finance Ministry's reaction shockingly lacks any connection to the realities of business life."

Let's remember what the whole discussion was originally all about: The sum Manfred Gentz was complaining about with regard to double taxation amounts to just $96 million, according to the company's calculations—a sum that Daimler's supervisory board chairman, Hilmar Kopper, could really refer to as "peanuts."

Klaus Köster, chief business editor of the *Stuttgarter Nachrichten*, gets straight to the point regarding the overreaction on Daimler's top floor: "That was not even 2 percent of 1998's annual profit, and only 0.07 percent of total sales." Köster correctly assesses that "such wide-reaching consequences as moving the company's headquarters and cutting jobs cannot be justified" on these grounds. Gentz, he says, gave "the impression that Daimler was using its importance to the region to put pressure on the political system."

The tax skirmishes aside, in mid-1999 it was still not clear whether, in the end, the amalgamation of these two large, completely different companies would actually be successful.

There was no question that the balance books were healthier than ever, with one record following another. But a very different answer was forthcoming with regard to corporate identity. The five-to-three German majority in the integration council spoke for itself, and there was a lot of trouble brewing behind the scenes. While the U.S. side has been blaming the Germans for their dominance and self-absorption, the latter have been saying that the Chrysler representatives are interested only in doing less and less work and making more and more profit. The beseeching words of the two CEOs Schrempp and Eaton have had little effect on this.

From the point of view of these kinds of problems, Schrempp is sitting on a powder keg. The world will blow up in Schrempp's face if the customers vote with their feet. If the board members continue to allow themselves the luxury of not bothering to define exactly where the company stands on important issues, then the question of whether the company is truly DaimlerChrysler or simply Daimler versus Chrysler will rear its ugly head. And if it turns into an intracompany battle of Daimler against Chrysler, the consequences will be most damaging, especially given the current climate of ruthless battle for commercial survival.

Asian Marketplace

We choose to see the opportunities.
Schrempp on the Asian financial crisis

In the mid-1990s Daimler's board of management saw good opportunities for expansion in China, Japan, and the other Asian countries. "This is the fastest-growing market," Eckhard Cordes explained at the time, and noted that the commercial vehicles sector in particular could expect impressive rates of growth: "Asia accounts for fifty percent of the commercial vehicle market and our share is 1 percent." As considerable rates of growth were being predicted for the Asian market, Cordes aggressively declared that "Asia is one of the main areas of our growth." By the year 2005, according to his target, Daimler sales there should rise to a sensational $21 billion. Achieving this aim would, however, prove impossible through the company's own growth alone, he said, so further takeovers could be expected in years to come.

But in the fall of 1997 stock exchanges throughout Asia were shaken by insecurity, threatening Daimler's ambitious plans. Divisional director Peter Fietzek nevertheless bravely waved the flag of the righteous: There was "nothing to change" in terms of company strategy, at least not for the moment. He added hopefully, "This turbulence does not mean the end of the Asian vision."

Daimler's chairman was not easily shaken. It took more than a crisis in Asia's economy, or the critical questions journalists were raising about his globalization process, to rattle Schrempp. Naturally, he said, the current business and currency crisis "will lead to short-term reductions in the growth rates of important Asian countries." And "various factors indicate that this setback" would be "overcome in the medium term." The region as a whole had "potential for above-average medium- and long-term growth in our opinion," which strengthened Schrempp's intention "to considerably expand our involvement in Asia."

Despite developments in Asia, Schrempp held firm to his strategy. "We hope," said the company chairman, discussing the strategic aims of Daimler-Benz AG, "to achieve them sooner." He, like Cordes, eventually wanted to see at least 20 to 25 percent of total sales coming from Asia.

Schrempp was acting like a global player: He considered the internationalization of the company ultimately essential to protect the company's investments in Germany. It was with this in mind that he made billion-dollar investments, built new factories, and opened up new markets for the company's products.

A minimum profit of 12 percent on capital investments was guaranteed even in those regions affected by the crisis in Asia, as Schrempp's philosophy demands positive thinking in all situations. Of course he really could not "guarantee" an increase in profits.

And if at the moment the sales market was not what it should be in Asia, then that must mean that opportunities were available elsewhere. "The number of automobiles that we cannot today distribute in Asia," said Schrempp, bursting with self-confidence, "would be very welcome on other markets." Only someone whose products were among the most highly desired, and whose customers were among the most patient, could make such a statement.

Schrempp was clearly more careful about what he said in internal discussions, as there were too many unanswered questions regarding the Asian problem. "The question of how to deal with Asia," acknowledged the chairman, "has still not been resolved." This cautious approach also appeared to be the more realistic position, for while 1997's automobile sales in Japan were stable compared to the year before, in the Far East as a whole sales actually fell by 7 percent. And the commercial vehicle divisions could only report a continuance of their previous poor performance in Asia.

Yet again Schrempp could not pass up the chance to criticize the visionary ideas of his predecessor. "I do not believe," said Schrempp, referring once again to Reuter, "that the old approach is ideal for the partnership."

During the first half of the nineties, under Reuter's control, Daimler carried out a number of negotiations with Mitsubishi aimed at forging a close alliance. "For a company like Daimler-Benz," said Reuter at the time, "a fundamental partnership with such a large group of companies would be a significant historical turning point." However, most of the development projects never came to fruition, and since then the Japanese automobile companies have come to face a crisis of major proportions.

The importance of the Asian market quickly brings Schrempp to the point: "Three billion people live there, buying power has grown rapidly, and they still have a lot of work to do on infrastructure." And in the spring of 1998, while Jürgen Schrempp and Robert Eaton were getting down to work on the DaimlerChrysler merger, Schrempp was already putting together his next takeover: "We have been holding talks with Nissan Diesel for quite a while," Daimler's chairman announced at the shareholders' general meeting in the summer of that year.

The aim was cooperation with Nissan's commercial vehicles division. And yet again Eckhard Cordes was in the thick of battle: "We want to invest expertise and money" to conquer the Asian market, he promised.

A few months later, DaimlerChrysler's chairman announced that if there was to be any capital participation in Nissan, "this would be completed before the end of the year."

When Schrempp and Eaton met Yoshikazu Hanawa, head of Nissan, in January 1999, Schrempp mentioned that there had been "constructive talks." On March 9 DaimlerChrysler's vehicles director, Tom Stallkamp, summed up hopefully, "We are still holding discussions." It seemed to be just a matter of hours before CEO Schrempp would announce the next coup and strengthen his claim on a world-leading position. Days later

Nissan's boss received the top Daimler team of Eckhard Cordes and Jür-gen Schrempp in the Japanese capital to finalize the engagement.

After the months of intensive bargaining, the collapse of the negoti-ations came as a surprise to many. Nissan Motor's lead negotiators pro-vided an unadorned view of why the talks collapsed: For one thing, Bob Eaton's ideas on gaining effective control over Nissan's management caused considerable ill will. Second, Schrempp linked the offer of a cap-ital infusion with his demand for corporate leadership—and in the process mistakenly relied on the supposed weakness of Nissan, deeply in debt to the tune of around $31 billion. This did not, however, explain why only two weeks later the French automobile company Renault was able to get the last laugh so easily and so quickly.

Jürgen Schrempp had played for high stakes and in the end had to make a total retreat—luckily, one would like to think, for under Euro-pean law Nissan would have no alternative but to file for bankruptcy. At the present time, no other vehicle producer has such an enormous level of debt as Nissan Motor—which surely did not become obvious to Schrempp and Eaton for the first time in March 1999.

The subsequent rhetorical conclusion to the Nissan matter consisted principally of empty statements: "We discussed the options openly and in a friendly atmosphere," said Schrempp in retrospect. Of course the Group was sticking to its aim of taking advantage of the growth oppor-tunities in Asia. However, "the integration of Daimler and Chrysler is our top priority." Surely Schrempp could have concentrated on this right from the start.

Daimler-Benz lacked any recognizable growth strategy for the Asian-Pacific area for many years. And though the projections were sunny, the fact was that Daimler's presence had been shrinking in all but Singapore and Taiwan. Now the failed association with Nissan had again set back their ambitious plans.

What strategy was Schrempp following when, at the shareholders' general meeting in May 1999, in nearly the same words as Dr. Cordes's, he put forth the theory that Asia was—"even after Nissan's refusal"—a "main area of our growth"? And what remained of Cordes's promises? It is true that Asia is the most rapidly growing market—though for the moment without DaimlerChrysler AG.

"Nkosi Sikele' i'Afrika"

I want to do something for the underprivileged now.
World Bank President James D. Wolfensohn, to Jürgen E. Schrempp

Southern Africa must be put on the agenda again.
Jürgen E. Schrempp, Chairman of the SAFRI initiative

They have known each other since way back in the eighties, a time during which investment banker James D. Wolfensohn advised a Mercedes-Benz subsidiary in the United States and Daimler's associated companies in other countries. "I heard his voice on the phone. He told me who he was and that he needed some advice," the American recalls of his very first conversation with Jürgen Schrempp.

Since then Schrempp and the Australian-born Wolfensohn have remained in contact.

"Jim has a lot of experience of restructuring, for example at Chrysler," Schrempp said of his continued consultations with Wolfensohn. This approach is typical of the current chairman of Daimler-Benz AG, who phones "ten to fifteen people on weekends," in particular "people who are not involved [directly in Schrempp's business dealings, and are therefore objective]. This gives me an unbelievable amount of input." And Wolfensohn is among those who have "provided lots of valuable advice."

Wolfensohn was one of those Schrempp turned to for advice as chairman of the Southern African Initiative (SAFRI), founded in April 1996 with the aim of improving the "economic and political framework" for cooperation between German business and the countries of southern Africa. Though the German business associations BDI and DIHT (German trade and industrial associations) and the Afrika-Verein e.V (a nonprofit NGO) support SAFRI, the interests of German business are represented though the "personal and active involvement" of executives such as Gerhard Cromme of Krupp, Jürgen Dormann of Hoechst, Martin Kohlhaussen of Commerzbank, Heinrich von Pierer of Siemens,

Bernd Pischetsrieder of BMW, and Jürgen Weber of Lufthansa, who are all on the executive committee.

So it was not surprising that Wolfensohn called Schrempp to ask his opinion on whether the banker should accept the presidency of the World Bank. As attractive as the offer might seem, Schrempp's initial question was "Why would you want to do that to yourself?" For the top position at the development bank would demand total involvement and leave no time for anything else.

"I have had a lot of success in my life," Jim Wolfensohn explains, "and I am doing very well materially." Wolfensohn, who became a U.S. citizen in 1980, could indeed look back on an above-average career. After working as a lawyer in Australia, he became a banker in Great Britain and then the United States, finishing up as chairman of an investment company in New York. There he founded his own investment bank in 1981, James D. Wolfensohn Inc., which has proved highly successful since then. "Now it was time," as Wolfensohn told Schrempp, "to give something back to the world."

Both men agree that life consists of more than just operating profit and shareholder value. That is why Schrempp says that "all of my heart is in" SAFRI, which he sees as his "personal contribution toward Africa." So in answer to Wolfensohn's question, Schrempp's final opinion was, "Then you must accept the job."

So in March 1995 Wolfensohn replaced Lewis Preston, who was ill, as the new president of the World Bank. Daimler's chairman admires his friend's unflagging enthusiasm: "It is incredible how he fights for the underprivileged. He's a real man of action."

Since Wolfensohn took over, there have been many changes at the World Bank, which used to be better known for its support of mammoth projects of questionable ecological and social benefit than for actually providing help to the poorest of the poor.

Schrempp is particularly impressed by the fact that with Wolfensohn "every top man [has to] get down to basics." And the basics of the World Bank are, for example, the slums of India.

This suggests that Schrempp and Wolfensohn share the same goals for global advancement.

In October 1997 in Cologne, Deutsche Welle (Germany's equivalent of the Voice of America or BBC's World Service) organized a "Southern African Forum." The day's main speaker, Jürgen Schrempp, spoke of regret that in Germany one had "long [talked] of a 'lost continent'" and that "'aid fatigue' was spreading throughout France." Schrempp said that "increasingly positive political and economic developments are taking place in the fourteen independent countries of southern Africa" that formed the Southern African Development Community (SADC). Even the exceptions proved the rule that "democracy and a market economy are increasingly asserting themselves."

Schrempp himself embodies the opportunities. "When Mr. Schrempp hears the words 'South Africa' he suddenly gets excited," says Josef C. Gorgels, manager of the SAFRI office at Daimler-Benz headquarters in Germany. As Gorgels points out, it was most important to push investment by medium-sized countries, for the large businesses "have already been present in South Africa for decades."

Indeed, Schrempp praised Claas Daun, chairman of the supervisory board of KAP, which has created eleven thousand jobs in the region.

Nongovernmental organizations concerned with development, such as DEAB in Stuttgart, the umbrella organization for developmental action groups, or Werkstatt Ökonomie, in Heidelberg, express a certain skepticism about the aims of the SAFRI initiative. They believe that SAFRI was "a body specially created" in order to influence business relations between Germany and the SADC countries "in the interests of [Germany's] large businesses." Ultimately, it was all about "developing a leading position on the South African market." German businesses want "a free-trade agreement between the EU and South Africa without protectionism" to make the export "of capital goods to South Africa easier." The accusation underlying their comments was that SAFRI was an initiative primarily directed toward promoting business interests—though Schrempp doesn't see it that way at all.

In November 1997 Schrempp hosted a joint World Bank–SAFRI conference in Stuttgart on the topic of private business and southern Africa. Most attention, however, was directed toward Jim Wolfensohn's remarks.

"I have seen the slums," he reported in a hoarse voice, and added, "This has changed my life completely." Anyone who has seen the settlements of those who do not know how they are going to survive the next day, he said, "has a reason to live." "Since then the purpose of [my] life has become focusing attention onto poverty." Ultimately, "there are not really two worlds. There is only one world.

"I represent an influential bank. But we need your support and we would like to support you, too," Wolfensohn wound up, to thunderous applause. And although the balancing act between the luxury meeting room at Stuttgart's Maritim Hotel, where the highest-ranking representatives of German business would soon indulge in a gourmet lunch, and the slums of the poorest conurbations of the world was a difficult one, Wolfensohn impressed his audience. "He was good, he was a world-shaker," said Christoph Köpke on his way out. Schrempp commented that Wolfensohn's enthusiasm "was infectious," and he was right.

Not everyone is as optimistic as Wolfensohn and Schrempp. Aside from all the fine words, Claas Daun, for one, has his doubts: "There is a large question mark about direct investment," says the businessman, who considers "SAFRI's chances very limited." The effect in the media may be positive, but "Mr. Schrempp's euphoria comes from the politician in him."

That same month members of German and South African high society gathered for a banquet at Düsseldorf's Steigenberger Parkhotel. All were in good spirits, including Schrempp, who was present in his capacity as honorary consul to South Africa, and his wife, Renate.

Matthias Kleinert, the Group's spokesman and like Schrempp a passionate singer, approached the microphone and announced, "Listen, Jürgen!" Kleinert had gathered ten dark-skinned waiters around him, and the group began a song that made Schrempp's half-African heart beat faster: "Nkosi Sikele' i'Afrika," South Africa's national anthem.

The African National Congress long ago made "Nkosi Sikele' i'Afrika" the anthem of the blacks. This song, written about a hundred years ago by the Methodist missionary and teacher Enoch Sontonga, gained a special position during apartheid. Indeed, it accompanied many blacks to their deaths during the time of South Africa's racist regime.

Following "Nkosi Sikele' i'Afrika" Schrempp and the others at the banquet started in on "Lord Bless Africa." Kleinert especially loves this song, and it has a deep significance for Schrempp as well, who admires the present government's courage in retaining the "classically white national anthem of the pre-Mandela era" as the second national anthem in addition to the blacks' song of freedom, "Nkosi Sikele' i'Afrika." "This was something really great psychologically, that the South Africans have two anthems—a demonstration of the readiness to belong together."

But many white South Africans do not see it that way today. "The country is losing a lot of know-how at the moment because the whites are leaving," says Daimler's chairman, who hopes that this development is only a reflection of the current mood and that many of them will return.

More important even than economic development for southern Africa is the need to encourage people to do away with Africa's negative image and give the people of Africa hope for a better future. The rest should be carried out by the Africans themselves, as far as this is possible.

Thus the greatest opportunity may well lie in the confidence and optimism spread by people like Schrempp. "It is not enough to complain," says Schrempp. "Everyone should do what they can within the framework of their capabilities. And I am happy to admit that I have greater opportunities than many other people."

And what about the accusation that Mercedes-Benz was merely using the framework of globalization to gain a greater market? Schrempp dismisses this idea by referring to the reality: "That is, at most, the case in South Africa." But the company is already well positioned there. The sales market in the other countries of southern Africa would be too small for years to come.

For Schrempp, though, the connection with South Africa clearly goes beyond business.

He says, "When I retire I will spend half the time in Europe and the other half year in South Africa." He feels at home there and can easily imagine the shape his life would take in Cape Town and on the farm in Eastern Transvaal.

Perhaps this love of the continent and its peoples explains the friendliness with which the manager is treated in South Africa, and to some extent elsewhere on the continent. "Nobody in the South African business world is as popular as Jürgen Schrempp," confirms Cape Town's Hugh Murray.

This was exactly the impression that Nelson Mandela also gave, on a political level, in January 1999 at the German Media Prize awards—an otherwise abstruse invention for the high society of Baden-Baden. After Lindiwe Mabuza, South Africa's ambassador, presented Jürgen Schrempp with the Order of Good Hope, South Africa's highest award for foreign citizens "who have represented the interests of South Africa in an outstanding manner," the winner of the Nobel peace prize took advantage of the opportunity to thank the German in the name of "the inhabitants of South Africa."

Schrempp, for his part, said of the most famous of all South Africans: "If there has ever been an ineradicable symbol of freedom other than the Statue of Liberty, then it is you, Mr. President." That, despite the pathetic gushing, the event did not degenerate into a boring gathering of the cognoscenti was solely thanks to Mandela himself, who toasted the public and joked,

"If you applaud my wife longer than me, I won't bring her with me next time."

It seems to be a foregone conclusion that after Schrempp's main business career comes to an end, his path will lead him back to South Africa to continue his private life there—even if in very different domestic company than originally planned.

The Job or the Marriage

The liaison with his head of staff has met with
embarrassed silence within the automobile company.
From an article on Jürgen E. Schrempp in the Wirtschaftswoche

Yes, this is a serious relationship.
Jürgen E. Schrempp

Below the article about how a laser-guided NATO bomb had wound up in the Belgrade bedroom of Serbian dictator Slobodan Milosevic, the

April 23, 1999, edition of German's most popular tabloid, *Bild*, carried the story of Jürgen Schrempp's separation from his wife, Renate. "She wanted me to start slowing down," the Lord of the Stars was quoted as saying, but "I wanted the merger with Chrysler."

Schrempp apparently felt that he had been presented with an impossible choice—"the job or the marriage." Given these circumstances, who could not understand that separation from his wife was unavoidable? "The challenge of my new duties means more to me than anything else in the world," he told the paper.

The story would simmer a little longer, as the article also mentioned the name of Schrempp's office manager, Lydia Deininger.

The next day's *Bild* carried a friendly profile of Deininger, describing her apartment, her animated chats on the phone, her fondness for teddy bears. The overall judgment was that she is "a competent, charming woman. Trustworthy. And she has flair." There is no doubt: Who could resist a woman like Lydia Deininger, especially after working with her over a period of years?

The message *Bild* was sending with this story was this: The poor, overworked top manager had sent his backward-thinking wife packing and taken up with his more understanding office manager—and *Bild* would give him absolution. This was perhaps a surprise, considering that the popular daily has a reputation for its "killer" stories, and that any politician crossing the editorial staff was likely to be attacked not just in a destructive front-page editorial but in articles detailing any evidence of adultery and affairs the tabloid could dig up. The list of those who have had their reputations ruined as a result of this Hamburg publication would fill volumes. Not for nothing was the *Bild* considered Germany's principal scandalmonger.

Conversely, the Lord of the Stars has long been afforded protection by Axel Springer's publishing house. Despite—or because of—his brusque and ruthless business decisions, and in spite of his obvious personal failings, the otherwise so dreaded daily has shielded Jürgen Schrempp. He shows his gratitude in his own way, with exclusive interviews and important announcements that boost circulation.

And so it came as no surprise that a potential *Bild* headline such as "Germany's Top Manager Cheats on Wife" was absent. The Hamburg paper instead assigned the blame, if there was any blame at all to be apportioned, to the morally neutral professional situation of the constantly stressed top manager: "The job destroyed his marriage." And "yet again a German management marriage has fallen foul of the 18-hour day and the jetting round the world."

In reply to the question posed by *Bild* reporter Jürgen W. Meyer about the relationship with his chief of staff, Daimler's chairman confirmed that he and Lydia have "a serious relationship"—a reply that is doubtless also applicable to his relationship with the Hamburg tabloid.

The *Bild* went on to say that Schrempp's "American Chrysler colleagues approve" of the liaison, and quoted an anonymous Chrysler employee enthusing about Schrempp's new partner. The whole company was behind Daimler's CEO and his beloved; at least this was the impression given by the half-page of gushing prose.

But a few conversations with those near to the chairman of the board are enough to ascertain that the perception within the Group was quite different. Schrempp's former wife, Renate, still has the best of reputations there. And not for nothing: She bore the brunt of her ex-husband's idiosyncrasies and escapades for so many years, accompanied her ex-husband through the difficult times of his defeats and through the low points, and gave him fresh heart. Thanks to her, and only her, did Schrempp remain chairman of the board of Daimler-Benz AG and thus become head of DaimlerChrysler. And, in what is certainly a considerable achievement, she remained what she had been three and a half decades earlier: a woman with both feet on the ground.

In contrast to *Bild*, Germany's business press commented critically on Schrempp's relationship with Deininger. The liaison has met "with embarrassed silence," *Wirtschaftswoche* informed us, adding that in the United States 70 percent of companies expressly forbid "a relationship between superiors and their subordinates" and that two-thirds of U.S. companies "react to office affairs uncompromisingly," with responses ranging from relocations right up to firings. When this business magazine inquired about the treatment of such cases at Chrysler in Detroit, they were met with "embarrassed silence from the otherwise so informative staff" there.

They have become careful at DaimlerChrysler's German headquarters. Lydia Deininger is now officially recognized as what she has already been, hidden away, for many years: the most influential woman in the company. As chief of staff, the well-paid Deininger (apparently Daimler-Chrysler pays her a salary of well over $133,000) manages Schrempp's daily business—from his personal correspondence and "preparation of documents for the management of the chairman of the board's office and the planning team" to his personal appearances.

Nevertheless, *Bunte,* a weekly magazine usually better known for being biased toward the Daimler chairman, describes Jürgen Schrempp in plain words: "A man addicted to work, success, and power."

AFTERTHOUGHTS

Cowards, Moral Cowards, and Some Heroes

Every day twenty people try to suck up to him.
Christoph Köpke, Chairman of MBSA,
on Schrempp's milieu in Germany

That not everyone I spoke to was prepared to answer in detail all the questions posed by the author of a book about the powerful cochairman of DaimlerChrysler AG is quite natural. However, I was lucky enough to meet many people, in Germany and particularly in South Africa, who were astonishingly open, refreshingly honest, and admirably courageous about reporting on their personal contact with Jürgen Schrempp, their various experiences with him, and his strengths and weaknesses as both man and manager. Many of their analyses and statements have been included in this book. In the case of some of those comments critical of Schrempp in tone or content, I have deliberately not attributed them to any particular person—even if anonymity was not requested. This approach was taken not because of any mistrust regarding Schrempp's reaction but to prevent repercussions, in whatever form, from other decision-makers at Daimler-Benz AG.

And while thoroughly critical reservations about Schrempp as a manager or as a person have been included in this outline of his career, statements that could have been highly injurious personally have been omitted.

In addition to the many interview subjects with backbone, I also came across some whose attitude I found particularly shocking.

In one office on a high floor with an unobstructed view of the Indian Ocean, for example, my wife and I listened for hours to a nervous-sounding individual talk about Schrempp's human merits and clever decisions, and the resulting career advances that logically followed.

I have the greatest understanding for the fact that not all those who made sometimes harsh statements wanted to be identified, but not for those who describe Schrempp as a superstar, megamanager, and near-god rolled into one and then tell me they have nothing more to say. Jürgen Schrempp is not godlike, and he would not claim to be, either. He has no need of such excesses, for his great human qualities and professional abilities are well known.

After an afternoon spent talking, that evening the dam broke, and the South African I had been interviewing raged that it was "incomprehensible" that people like Schrempp and Stöckl could become members of the Daimler board. "Where would German business be if it put such men in control?" was his provocative question. The discussion climaxed in a fierce dispute about the human and professional qualities of Daimler's management: "Don't you Germans have anything better to offer than Schrempp and Stöckl?" was one of the vehement accusations.

I suddenly found myself playing devil's advocate, trying to present a contrasting position, and so my wife and I decided to bid farewell to this critic of Schrempp's career for the evening.

The next morning all was forgotten; my subject again told me that Schrempp was the superman, the perfect manager. I was intensely annoyed, particularly as I knew that I was hardly likely to get another opportunity to talk for this man again personally. I had hoped that that day's meeting would provide me with background information to the rather superficial accusations made the evening before. I could sense, from a number of extremely interesting allusions, that this man knew a lot about Schrempp—positive as well as negative.

What had happened in the few hours since our last meeting? Had the intimate atmosphere of the previous evening encouraged him to reveal what he saw in the cold light of morning as too much? Had the man suddenly realized that he was talking to an author, and that afterward everything would be there in black and white for all to see?

He had had a sleepless night, the South African admitted when I pointed out the obvious discrepancy of his statements. The man was a business insider and a longtime friend of Schrempp, and had boasted about having spent "many evenings and parties together."

I finally threw all caution to the wind and asked him directly what economic success Schrempp had achieved in South Africa. "Schrempp was not successful; he just presented the figures in the best possible light" was the clear answer. "He achieved personal success despite the financial losses. His contact with the board was helpful," added my interviewee, who—please note—still has a contract with Mercedes-Benz AG. But apparently even this was still too negative, for he immediately added that I must on no account quote him. Nevertheless I was hopeful of learning more. In the meantime, however, I had to agree with his characterization of himself, as he again and again reproached himself: "You know, I'm a coward!"

So it was left to critics such as Martine Dornier-Tiefenthaler to tell it like it is. "My job doesn't depend on him," she said. "I can see him as he is. Which is why he so easily gets nervous in my presence."

Unfortunately only a few people who know him well, close friends, have the courage to tell Schrempp the truth to his face. "I have the most stormy relationship with him," boasted Hugh Murray in Cape Town. This is mainly because "I tell him when he has done something stupid." And that is precisely what is so unusual and, in Hugh Murray's case, so easy to explain: In contrast to the more than 300,000 employees dependent on Daimler-Benz AG for their livings, the Cape Town publisher can express his criticism without having to fear any form of repression whatsoever. When Murray says that some of those around Schrempp lie, then the question of why they lie automatically arises.

Today's chairman of Daimler has repeatedly emphasized that only constructive criticism is useful. This man believes in the power of the better argument—and is convinced that the better arguments are almost always his. This sounds arrogant, and sometimes it is. However, it must be said that under his leadership there is now a far more open and clearly more communicative atmosphere at Daimler-Benz than under his predecessors, and this is confirmed by members of the board of management and the supervisory board.

The red carpet is rolled out when the chairman of the board visits one of his factories, and it may be that Hollywood-style facades are even erected. One must be understanding about this, too. For why should employees voluntarily risk telling a powerful person, such as Schrempp undoubtedly is, the truth about any mismanagement or deplorable state of affairs that may exist?

Near Munich I met a former top manager of Dasa with long experience of the structural initiatives that have been undertaken over the years, and close personal knowledge of how Schrempp failed in this or that endeavor. At the end I asked him (as I have asked all those I interviewed) whether he had made any statements that I could not ascribe to him, and he replied, "On no account are you to mention my name." It was not difficult to find out why: In years gone by he raised criticisms within Dasa itself and had been given a warning for it. One of his sons is now successfully employed in the company. "I have been informed that my son's career will not be damaged," said the former top manager, "if I do not make any negative statements."

Ultimately two aspects in particular explain why Jürgen Schrempp is still, according to the accusations from within his own company, surrounded by moral cowards. For one, Schrempp's system was completely oriented toward operating profit, shareholder value, and high returns, in the same style as that of his American model, Jack Welch. The logical consequence of this is the optimization of work procedures and a ruthless restructuring process. As a result, guarantees can be of limited duration at best, explaining the fear many employees have about layoffs and the possibility of losing their livelihoods. Furthermore, the ruthlessly hard system promotes, if not demands, a conformist mentality.

During his career Schrempp has not exactly earned a reputation as an apostle of peace or a compassionate Samaritan. Thus the chairman of the board must ask himself what he has done to bring about this situation. "One needs to be careful," said one former confidant, for "he can be resentful." And then, according to this individual, "it took years till he forgave." His dry final comment: "That's really great!"

It is doubtful whether virtues such as honesty and the ability to be usefully critical will have an easy time of it at DaimlerChrysler in years to come. If the fears of Daimler unionists such as Gerhard Zambelli or Gerd Rathgeb are well founded, then the merger with Chrysler will indeed lead to a considerable loss of jobs. The Chrysler Corporation made conspicuously higher net profits than Daimler-Benz AG but had only slightly more than one-third of the employees.

So what will happen when American ideas of profit and returns are superimposed, even partially, onto the classy German company? Who will then dare to register discontent as an employee of DaimlerChrysler AG?

It can only be hoped that the number of moral cowards in Germany will decline and that the number of the brave, especially at the German headquarters, rises again. Schrempp has the ability to instill courage in others and is himself courageous. His style, his openness, and his optimism are encouraging. His agreement to this book project is not the least of the examples of his courage—and his way of showing backbone should be a spur to others.

INDEX

About the Author

Jürgen Grässlin is the Speaker of KAD (Organization of Critical Shareholders of Daimler-Benz) and has been one of Schrempp's most outspoken critics. Schrempp nonetheless agreed to grant five interviews with the author for this book. Grässlin lives in Frieburg, Germany.